I0499774

Printed in the United States of America

M A T H

COMPETITION QUESTIONS
for
4th & 5th GRADE.

2000 Questions with Answers

LEVEL - 1

Published by Creatspace: 06/04/2020

Book design and interior formatting by Sahib Kazimov
e-mail: kazimli@hotmail.com

ISBN-13 : 978-1717399915
ISBN-10: 1717399916

Tayyip oral

Veysel Dereli

M A T H

COMPETITION QUESTIONS
for
4th & 5th GRADE

2000 Questions with Answers

LEVEL - 1

555 Math Book series

1) 1000 Logic & Reasoning Questions for Gifted and Talented Elementary School Students

2) 555 SAT Math(555 Questions with Solution)

3) 555 GEOMETRY (555 Questions with Solution)

4) 555 GEOMETRY Problems for High School Students

5) 555 ACT Math (555 Questions with Solution)

6) 555 ACT Math (555 Questions with Answers)

7) 555 ADVANCED Math Problems - for Middle School Students

8) 555 MATH IQ Questions for High School Students

9) 555 MATH IQ Questions for Middle School Students

10) 555 MATH IQ Questions for Elementary School Students

11) 555 GEOMETRY Formula handbook for SAT, ACT, GRE

12) GEOMETRY Formula Handbook

13) ALGEBRA Handbook for Middle School Students

14) GEOMETRY for SAT&ACT (555 Questions with Answers)

15) 555 Gifted and Talented for Middle School Students

16) ALGEBRA for the New SAT (1111 Questions with Answers)

17) ALGEBRA for the ACT (1080 Questions with Answers)

18) 555 MATH IQ for Elementary School Students (Second Edition)

19) TSI MATH (Texas Success Initiative)

20) ACCUPLACER MATH PREP

21) CLEP College Algebra

22) MATH WORD Problems (540 Questions with Answers)

Table of Contents

Preface

Math Competition book is a developmental practice questions text for all students who are preparing for math contest. It uses 2000 practice questions. this book to develop and improve students practice skills.

Math Competition questions are a challenge for students in grade 4 and 5. This book level is one. Variety of challenge problems that include easy, medium, and hard math problem cover. In this book you see different questions. However math competition question books are great starting point to train students for math competition. This book is good for elementary school students who wants extra practice to prepare for math contest. This book includes 2000 is very much interested in doing the questions.

I hope you have been enjoyed these books..

Tayyip Oral

oral_tayyip@yahoo.com

tayyiporal@gmail.com

NUMBERS

Natural numbers: {1,2,3,4,…}
Whole numbers: {0,1,2,3,4,5,…}
Integer numbers: {…,-3,-2,-1,0,1,2,3,…}
Any element of the $\left\{\dfrac{a}{b}\right\}$ is called rational numbers: $\left\{\dfrac{7}{2}, \dfrac{1}{2}, 3, \dfrac{1}{4}, …\right\}$

TEST – 1

1. Which number is rational number?
   A) 7   B) $\dfrac{1}{3}$ C) $\dfrac{5}{3}$ D) All

4	-8	12	24	9
$-\dfrac{48}{16}$		$6^2 \div 6$		7
$\dfrac{1}{7}$		$2 \div 8$		$\dfrac{1}{9}$

2. Which number is whole number?
   A) $\dfrac{27}{3}$ B) $\dfrac{27}{7}$ C) $-\dfrac{24}{8}$ D) $\dfrac{6}{4}$

3. Which number is integer number?
   A) $\dfrac{8}{6}$ B) $-\dfrac{42}{7}$ C) $\dfrac{1}{3}$ D) $\dfrac{7}{2}$

7. Which number is integer number?
   A) -3^2 B) $-\dfrac{3}{9}$ C) $-4^2 \div 32$ D) $\dfrac{1}{4}$

Set A=$\left\{-4, -2, 1, 7, 18, \dfrac{144}{12}\right\}$

4. How many whole numbers are at set A?
   A) 5   B) 4   C) 3   D) 2

8. How many whole numbers are at the table?
   A) 6   B) 5   C) 4   D) 3

5. How many integer numbers at the set A?
   A) 7   B) 6   C) 5   D) 4

9. How many integer numbers are at the table?
   A) 9   B) 8   C) 6   D) 7

6. How many rational numbers at the set A?
   A) 7   B) 8   C) 5   D) 6

10. How many rational numbers are at the table?
    A) 11   B) 10   C) 9   D) 8

Prime numbers: {2,3,5,7,11,13,17,19,23,29,31,37,41,...}

Composite numbers: {4,6,8,9,10,12,14,15,16,18,...}

Even numbers: {...,-8,-6,-4,-2,0,2,4,6,8,...}

Odd numbers: {-7,-5,-3,-1,1,3,5,7,...}

TEST – 2

Set A={-6,-4,0,2,5,7,11,12,17,19,27,31}

1. Which numbers in the set A are prime numbers?
   A) 2,5,7,11,12,19
   B) 2,5,7,11,17,19
   C) 5,7,11,17,19,22
   D) 2,5,7,11,17,19,32

2. Which numbers in the set A are the composite numbers?
   A) 12,19,27,31
   B) 12,27
   C) 2,12,27,31
   D) 0,2,12,27

3. How many even integer numbers in the set A?
   A) 4   B) 5   C) 6   D) 7

4. How many prime numbers are between 10 to 50?
   A) 14   B) 15   C) 16   D) 11

5. How many composite numbers are between 11 to 31?
   A) 14   B) 15   C) 16   D) 17

6. Which number is prime number?
   A) 52   B) 62   C) 71   D) 93

7. Which number is composite number?
   A) 67   B) 77   C) 89   D) 97

8. Find the sum all prime numbers between 40 to 50.
   A) 121   B) 131   C) 141   D) 151

9. Find the sum all composite numbers between 31 to 41.
   A) 248   B) 297   C) 307   D) 317

10. Given the set A, {23, 31, 41, 53, 61, 71} list.
    A) Composite numbers
    B) Prime numbers and composite numbers
    C) Even numbers
    D) Only prime numbers

Prime numbers: {2,3,5,7,11,13,17,19,23,29,31,37,41,...}

Composite numbers: {4,6,8,9,10,12,14,15,16,18,...}

Even numbers: {...,-8,-6,-4,-2,0,2,4,6,8,...}

Odd numbers: {-7,-5,-3,-1,1,3,5,7,...}

TEST – 3

Set A:
{17,21,27,28,42,60,61,71,80,90,131,148}

1. How many odd numbers are at the set A?
   A) 7   B) 6   C) 5   D) 4

2. How many even numbers are at the set A?
   A) 8   B) 7   C) 6   D) 5

   Even numbers=E
   Odd numbers=O

3. E+E=...
   A) E   B) O   C) not there

4. Odd+odd=...
   A) E   B) O   C) not there

5. E+E+E=...
   A) E   B) O   C) not there

6. Odd+odd+odd=...
   A) O   B) E   C) not there

7. 2^5+3^9=...
   A) O   B) E   C) not there

8. Odd number - odd number + even number = ...
   A) O   B) E   C) not there

   E^n=E
   O^n=O

9. 5^{64}=...
   A) E   B) O   C) not there

10. 3^{81}=...
    A) O   B) E   C) not there

Some summative formulas

1) $1+2+3+4+\ldots+n=\dfrac{n(n+1)}{2}$

2) $2+4+6+8+\ldots+2n=n(n+1)$

3) $1^2+2^2+3^2+\ldots+n^2=\dfrac{n(n+1)\cdot(2n+1)}{6}$

4) $1^3+2^3+3^3+\ldots+n^3=\left(\dfrac{n(n+1)}{2}\right)^2$

TEST – 4

1. $1+2+3+\ldots+10=?$

   A) 65   B) 55   C) 45   D) 35

2. $1+2+3+\ldots+30=?$

   A) 465   B) 485   C) 495   D) 505

3. $2+4+6+\ldots+24=?$

   A) 156   B) 166   C) 176   D) 30

4. $2+4+6+\ldots+40=?$

   A) 400   B) 420   C) 430   D) 440

5. $1^2+2^2+3^2+\ldots+10^2=?$

   A) 280   B) 335   C) 375   D) 385

6. $1^2+2^2+3^2+\ldots+20^2=?$

   A) 2810   B) 2840   C) 2860   D) 2870

7. $12+13+14+\ldots+32=?$

   A) 217   B) 218   C) 462   D) 220

8. $16+18+20+\ldots+40=?$

   A) 364   B) 362   C) 360   D) 820

9. $1^3+2^3+3^3+\ldots+10^3=?$

   A) 3012   B) 3018   C) 3020   D) 3025

* $1+3+5+\ldots+(2n-1)=n^2$

10. $1+3+5+\ldots+19=?$

   A) 90   B) 96   C) 100   D) 110

GREATEST COMMON DIVISION (GCD)
GREATEST COMMON FACTOR (GCF)

TEST – 5

1. Find the greatest factor (GCF) of 12 and 30.
   A) 6   B) 8   C) 10   D) 15

2. Find the GCF of 15 and 25.
   A) 5   B) 10   C) 15   D) 25

3. Find the GCF of 12 and 34.
   A) 2   B) 4   C) 16   D) 17

4. Find the GCF of 10 and 70.
   A) 7   B) 15   C) 20   D) 10

LEAST COMMON MULTIPLY (LCM)

5. What is the Least Common of 12 and 48 or LCM (12; 48)
   A) 12   B) 24   C) 36   D) 48

6. What is the Least Common of 16 and 15?
   A) 120   B) 180   C) 190   D) 240

7. Find the LCM(5;30).
   A) 20   B) 30   C) 35   D) 40

8. Find the LCM(6;45).
   A) 90   B) 70   C) 60   D) 45

9. Find the LCM(15;25).
   A) 45   B) 65   C) 75   D) 100

10. Find the GCF of 18 and 24.
    A) 9   B) 8   C) 7   D) 6

EXPONENTS

4 - EXPONENT
3 - BASE

In the expression 3^4, 3 is called the base and 4 is called the exponent.

- 5^2: 5 to the second power
- 4^3: 4 cubed or 4 to the third power
- 3^6: 3 to the sixth power
- $\left(\dfrac{3}{5}\right)^4$: $\dfrac{3}{5}$ to the fourth power.

TEST – 6

1. $2^3 + 3^2 = ?$
   A) 12   B) 17   C) 16   D) 18

2. $4^3 + 3^4 = ?$
   A) 24   B) 145   C) 155   D) 160

3. $9^2 - 8^2 = ?$
   A) 1   B) 9   C) 8   D) 17

4. $(36 \div 6)^2 = ?$
   A) 36   B) 30   C) 25   D) 24

5. $\left(\dfrac{2}{5}\right)^3 = ?$
   A) $\dfrac{8}{15}$ B) $\dfrac{6}{25}$ C) $\dfrac{8}{125}$ D) $\dfrac{8}{25}$

6. $(24 \div 4 + 4)^2 = ?$
   A) 70   B) 80   C) 90   D) 100

7. $(7 \div 7 + 7)^2 = ?$
   A) 64   B) 81   C) 100   D) 49

8. Find the $\dfrac{2}{3}$ to the third power.
   A) $\dfrac{8}{9}$ B) $\dfrac{8}{27}$ C) $\dfrac{4}{27}$ D) $\dfrac{6}{27}$

9. Find the $\dfrac{1}{2}$ to the cubed.
   A) $\dfrac{1}{2}$ B) $\dfrac{1}{4}$ C) $\dfrac{1}{8}$ D) $\dfrac{1}{16}$

10. $12^2 - 10^2 = ?$
    A) 12   B) 24   C) 44   D) 48

MULTIPLICATION AND DIVISION POSITIVE AND NEGATIVE NUMBERS

$$(+) \cdot (+) = + \qquad\qquad (+2) \cdot (+3) = +6$$
$$(+) \cdot (-) = - \qquad\qquad (+3) \cdot (-4) = -12$$
$$(-) \cdot (-) = + \qquad\qquad (-2) \cdot (-3) = +6$$
$$(-) \cdot (+) = - \qquad\qquad (-2) \cdot (+5) = -15$$
$$(-) \div (-) = + \qquad\qquad (-8) \div (-2) = +4$$
$$(-) \div (+) = - \qquad\qquad (-9) \div (+3) = -3$$
$$(+) \div (-) = - \qquad\qquad (+10) \div (-2) = -5$$
$$(+) \div (+) = + \qquad\qquad (+8) \div (+2) = +4$$

TEST – 7

1. $(-8) \cdot 7 = ?$
   A) 56   B) -56   C) -1   D) -7

2. $-24 \div (-3) = ?$
   A) -8   B) 8   C) $\dfrac{1}{8}$ D) $-\dfrac{1}{8}$

3. $(8-9) \cdot (9-8) = ?$
   A) 1   B) 2   C) 17   D) -1

4. $(10-8) \cdot (-8) = ?$
   A) -10   B) -12   C) -16   D) 16

5. $(-24 \div 4) \div (-6) = ?$
   A) 1   B) 2   C) -2   D) -1

6. $(-36 \div 36) \div (-36) = ?$
   A) 1   B) -1   C) $\dfrac{1}{36}$ D) $-\dfrac{1}{36}$

7. $A = -24 \div (-8)$, $B = -8 \div (-2)$
   A+B=?
   A) -7   B) 7   C) 8   D) -8

8. $A = 8 \cdot (-3)$, $B = -3 \cdot 8$
   A+B=?
   A) -48   B) 12   C) 24   D) 36

9. $\left(-\dfrac{4}{3}\right) \cdot (-18) = ?$
   A) 12   B) 24   C) -12   D) -24

10. $(-9) \div 9 + (-8+8) = ?$
    A) 2   B) -2   C) 0   D) -1

REMAINDER

$$\begin{array}{r} 29 \\ 5\overline{\smash{)}149} \\ -\underline{10} \\ 049 \\ -\underline{45} \\ 0\textcircled{4} \rightarrow remainder \end{array}$$

TEST – 8

1. What is the remainder when 665 is divided by 3?
   A) 0   B) 1   C) 2   D) 3

2. What is the remainder when 7787 is divided by 5?
   A) 0   B) 2   C) 3   D) 4

3. What is the remainder when 9999 is divided by 9?
   A) 0   B) 1   C) 2   D) 5

4. What is the remainder when 9999999 is divided by 5?
   A) 0   B) 1   C) 2   D) 4

5. What is the remainder when 252525 is divided by 25?
   A) 4   B) 3   C) 2   D) 0

6. What is the remainder when $(100 \div 5 + 5)$ is divided by 5?
   A) 4   B) 3   C) 2   D) 0

7. What is the remainder when six-thousand is divided by 25?
   A) 0   B) 1   C) 2   D) 3

8. What is the remainder when $(24 \div 4 + 4)^2$ is divided by $(8 \div 8 + 4)$?
   A) 0   B) 1   C) 2   D) 4

9. What is the remainder when 7777777 is divided by 3?
   A) 4   B) 3   C) 0   D) 1

10. What is the remainder when 999999999927 is divided by 3?
    A) 0   B) 1   C) 2   D) 3

WORD PROBLEM – DOZEN

TEST – 9

1. Luis buys 8 dozen pencil. How many pencil does he have in all?
   A) 24   B) 48   C) 72   D) 96

2. Ronaldo has 7 dozen books. How many books does he have?
   A) 49   B) 84   C) 94   D) 102

3. Anderson bought 5 dozen eggs. How many eggs does he have?
   A) 60   B) 54   C) 48   D) 30

4. Javier has 12 dozen duck in farm. How many ducks does he have?
   A) 144   B) 134   C) 124   D) 112

5. Luis and Tom have two hundred eighty-eight books. How many dozen books do they have in all?
   A) 21   B) 22   C) 24   D) 26

6. Luis have 72 marbles. How many dozen marbles did he have?
   A) 4   B) 5   C) 6   D) 7

7. Javier has 54 green marbles, Orlando has 84 blue marbles, Jack has 6 yellow marbles. How many dozen they have marbles?
   A) 11   B) 12   C) 13   D) 14

8. Susana bought 8 dozen small bread. How many small breads did he buy?
   A) 86   B) 96   C) 216   D) 112

9. Food store sold 11 dozen box water. How many waters did it sale?
   A) 121   B) 132   C) 142   D) 152

10. One soccer club has 15 dozen soccer balls. How many balls does it have?
    A) 144   B) 154   C) 180   D) 174

SIMPLIFY FRACTION (REDUCE)

$$\frac{3}{6} = \frac{1 \times \cancel{3}}{2 \times \cancel{3}} = \frac{1}{2} \qquad \frac{6}{20} = \frac{\cancel{2} \times 3}{\cancel{2} \times 2 \times 5} = \frac{3}{10} \qquad \frac{28}{38} = \frac{\cancel{2} \times 14}{\cancel{2} \times 19} = \frac{14}{19} \qquad \frac{35}{45} = \frac{\cancel{5} \times 7}{\cancel{5} \times 9} = \frac{7}{9}$$

TEST – 10

1. Find the simplify form of $\frac{14}{35}$.

   A) $\frac{2}{5}$ B) $\frac{2}{7}$ C) $\frac{7}{5}$ D) $\frac{5}{6}$

2. Find the simplify form of $\frac{36}{48}$.

   A) $\frac{6}{9}$ B) $\frac{4}{5}$ C) $\frac{4}{3}$ D) $\frac{3}{4}$

3. Find the simplify form of $\frac{84}{64}$.

   A) $\frac{16}{22}$ B) $\frac{16}{21}$ C) $\frac{21}{16}$ D) $\frac{23}{17}$

4. Find the simplify form of $\frac{225}{125}$.

   A) $\frac{9}{5}$ B) $\frac{5}{9}$ C) $\frac{10}{9}$ D) $\frac{9}{10}$

5. Find the simplify form of $\frac{36}{78}$.

   A) $\frac{18}{19}$ B) $\frac{19}{18}$ C) $\frac{6}{13}$ D) $\frac{17}{18}$

6. Find the simplify form of $\frac{35}{75}$.

   A) $\frac{5}{13}$ B) $\frac{5}{17}$ C) $\frac{7}{15}$ D) $\frac{7}{12}$

7. Find the simplify form of $\frac{36}{52}$.

   A) $\frac{9}{13}$ B) $\frac{9}{14}$ C) $\frac{8}{13}$ D) $\frac{4}{13}$

8. Find the simplify form of $\frac{84}{35}$.

   A) $\frac{12}{7}$ B) $\frac{7}{12}$ C) $\frac{12}{8}$ D) $\frac{12}{5}$

9. Find the simplify form of $\frac{85}{289}$.

   A) $\frac{5}{17}$ B) $\frac{6}{17}$ C) $\frac{5}{18}$ D) $\frac{17}{5}$

10. Find the simplify form of $\frac{112}{169}$.

   A) $\frac{13}{8}$ B) $\frac{8}{13}$ C) $\frac{7}{13}$ D) $\frac{14}{13}$

PRODUCT -(MULTIPLICATION)

Example: Find the product of 7 and 5.   *Solution:* 7×5=35
Example: Find the product of 12 and 6.   *Solution:* 12×6=72

TEST – 11

1. Find the product of 15 and 5.
   A) 20   B) 75   C) 85   D) 3

2. Find the product of 7 and 15.
   A) 102   B) 105   C) 110   D) 8

3. Find the product of $\dfrac{5}{11}$ and 121.
   A) 11   B) 22   C) 44   D) 55

4. Find the product of $\dfrac{3}{4}$ and 84.
   A) 53   B) 62   C) 63   D) 64

5. Find the product of 4 and 4.
   A) 12   B) 24   C) 16   D) 64

6. Find the product of 5,5 and 7.
   A) 145   B) 155   C) 165   D) 175

7. Find the product of 3^2 and 2^3.
   A) 72   B) 62   C) 52   D) 48

8. Find the product of 4^2 and 2^4.
   A) 246   B) 256   C) 266   D) 276

9. Find the product of $(99)^0$ and $(1)^{99}$.
   A) 99   B) 99×99   C) 1×1   D) 1×98

10. Find the product of 5^2 and 2^5.
    A) 600   B) 700   C) 800   D) 900

ROUND TO ESTIMA

- Rounding to the nearest 10.
  73 → 70   36 → 40
  44 → 40   67 → 70
  62 → 60   89 → 90

- Rounding to the nearest 100.
  536 → 500
  678 → 600
  425 → 400

- Rounding to the nearest 1000.
  3641 → 4000
  2917 → 3000
  4262 → 4000
  2375 → 2000

TEST – 12

1. Estimate: 33+43=?
   A) 60   B) 70   C) 80   D) 90

2. Estimate: 42+52=?
   A) 130   B) 140   C) 90   D) 160

3. Estimate: 98+88+78=?
   A) 240   B) 250   C) 260   D) 270

4. Estimate: 324+428+664
   A) 1300   B) 1200   C) 1100   D) 1400

5. Estimate: 774+664+444+333=?
   A) 2400   B) 2300   C) 2200   D) 2100

6. Round 6345 to the nearest ten.
   A) 6350   B) 6352   C) 6300   D) 6355

7. Round 7399 to the nearest ten.
   A) 7450   B) 7400   C) 7461   D) 7500

8. Round 9685 to the nearest hundred.
   A) 9780   B) 9690   C) 9700   D) 9860

9. Round 384 to the nearest hundred.
   A) 380   B) 400   C) 390   D) 391

10. Round 98687 to the nearest thousand.
    A) 99000   B) 99700
    C) 99600   D) 99800

DECIMAL MULTIPLICATION

$$0.a \times 0.b = \frac{a}{10} \times \frac{b}{10}$$

$$(0.a) \times (0.bc) = \frac{a}{10} \times \frac{bc}{100} = \frac{a \times bc}{1000}$$

$$0.2 \times 0.3 = \frac{2}{10} \times \frac{3}{10} = \frac{6}{100} = 0.06$$

$$0.3 \times 0.24 = \frac{3}{10} \times \frac{24}{100} = \frac{72}{1000} = 0.072$$

$$0.a = \frac{a}{10} \qquad 0.ab = \frac{ab}{100} \qquad 0.abc = \frac{abc}{1000}$$

TEST – 13

1. $0.9 \times 0.12 = ?$
   A) 1.08   B) 0.108   C) 0.109   D) 10.8

2. $0.6 \times 0.7 = ?$
   A) 42   B) 4.2   C) 0.42   D) 0.042

3. $0.12 \times 0.13 = ?$
   A) 15.6   B) 1.56
   C) 0.156   D) 0.0156

4. $0.7 \times 0.11 = ?$
   A) 0.77   B) 7.7   C) 0.077   D) 77

5. $1.2 \times 7.4 = ?$
   A) 88.8   B) 8.80   C) 8.88   D) 8.98

6. $1.6 \times 6.1 = ?$
   A) 97.6   B) 9.76   C) 8.76   D) 7.76

7. $0.9 \times 0.22 = ?$
   A) 1.98   B) 7.88   C) 0.198   D) 0.188

8. $0.77 \times 7.7 = ?$
   A) 5.919   B) 5.819   C) 5.929   D) 4.919

9. $0.012 \times 0.03 = ?$
   A) 0.036   B) 0.0036
   C) 0.00036   D) 0.36

10. $0.44 \times 0.55 = ?$
    A) 0.242   B) 0.0242
    C) 0.342   D) 0.0342

FIND THE NEXT TERM IN THE ARITHMETIC SEQUENCE

Sequence:

1st term 3rd term

4, 7, 10, 13, 16, ...

2nd term 4th term

4, 7, 10, 13, 16, ...
$+3$ $+3$ $+3$ $+3$

In general we could write an arithmetic sequence like this: {a, a+d, a+2d, a+3d, ...}
a: is the first term,   d: is the difference between the terms (common difference)
Example: sequence: 4,7,10,13,16,19,........ a=4, d=3

TEST – 14

1. The next term in the arithmetic sequence
    5,12,19,26,33,...is
    A) 39   B) 40   C) 42   D) 44

2. Find the 8th term.
    A) 52   B) 54   C) 56   D) 58

3. The next term in the arithmetic sequence
    {7,13,19,25,...} is
    A) 31   B) 32   C) 33   D) 34

4. Find the 7th term
    A) 41   B) 42   C) 43   D) 44

5. The next term in the arithmetic sequence
    {9,17,25,33,41,...} is
    A) 53   B) 52   C) 51   D) 49

6. {9,17,25,33,41,...)
    Find the common difference.
    A) 5   B) 6   C) 7   D) 8

7. The next term in the arithmetic sequence
    {3,12,21,30,39,...} is
    A) 46   B) 47   C) 48   D) 49

8. Find the common difference.
    A) 5   B) 6   C) 7   D) 9

9. The next term in the arithmetic sequence
    {4,15,26,37,...} is
    A) 48   B) 49   C) 50   D) 51

10. Find the 7th term.
    A) 64   B) 65   C) 66   D) 70

DECIMAL NUMBERS CONVERT TO FRACTION

1) $0.a = \dfrac{a}{10}$
$\qquad 0.3 = \dfrac{3}{10}$

2) $0.ab = \dfrac{ab}{100}$
$\qquad 0.25 = \dfrac{25}{100}$

3) $0.abc = \dfrac{abc}{1000}$
$\qquad 0.127 = \dfrac{127}{1000}$

TEST – 15

1. 0.16=…………(fraction)

A) $\dfrac{16}{10}$ B) $\dfrac{16}{100}$ C) $\dfrac{4}{100}$ D) $\dfrac{4}{10}$

2. 0.8=…………….. (fraction)

A) $\dfrac{8}{10}$ B) $\dfrac{8}{100}$ C) $\dfrac{2}{10}$ D) $\dfrac{4}{10}$

3. 0.24=……….. (fraction)

A) $\dfrac{6}{100}$ B) $\dfrac{100}{24}$ C) $\dfrac{24}{10}$ D) $\dfrac{24}{100}$

4. 0.10=?...........(fraction)

A) 1 B) $\dfrac{10}{10}$ C) $\dfrac{10}{100}$ D) $\dfrac{10}{1000}$

5. 0.35=?............(fraction)

A) $\dfrac{35}{10}$ B) $\dfrac{35}{100}$ C) $\dfrac{7}{50}$ D) $\dfrac{5}{20}$

6. 0.55=……… (fraction)

A) $\dfrac{5}{10}$ B) $\dfrac{35}{100}$ C) $\dfrac{11}{20}$ D) $\dfrac{22}{20}$

7. 0.18=…………(fraction)

A) $\dfrac{18}{10}$ B) $\dfrac{9}{25}$ C) $\dfrac{9}{50}$ D) $\dfrac{18}{50}$

8. 0.6=…………. (fraction)

A) $\dfrac{3}{5}$ B) $\dfrac{6}{100}$ C) $\dfrac{3}{10}$ D) $\dfrac{3}{100}$

9. 0.44=? (fraction)

A) $\dfrac{44}{10}$ B) $\dfrac{22}{10}$ C) $\dfrac{22}{50}$ D) $\dfrac{22}{100}$

10. 0.75=? (fraction)

A) $\dfrac{3}{4}$ B) $\dfrac{75}{10}$ C) $\dfrac{15}{25}$ D) $\dfrac{75}{1000}$

FRACTION ADDITION AND SUBSTRACTION

Same denominator: $\dfrac{x}{a} + \dfrac{y}{a} = \dfrac{x+y}{a}$ Example: $\dfrac{7}{11} + \dfrac{5}{11} = \dfrac{7+5}{11} = \dfrac{12}{11}$

Same denominator: $\dfrac{x}{a} - \dfrac{y}{a} = \dfrac{x-y}{a}$ Example: $\dfrac{9}{14} - \dfrac{5}{14} = \dfrac{9-5}{14} = \dfrac{4}{14}$

TEST – 16

1. $\dfrac{4}{12} + \dfrac{2}{12} = ?$

   A) $\dfrac{1}{2}$ B) $\dfrac{1}{3}$ C) $\dfrac{2}{12}$ D) $\dfrac{7}{12}$

2. $\dfrac{3}{11} + \dfrac{4}{11} = ?$

   A) $\dfrac{1}{11}$ B) $\dfrac{7}{11}$ C) $\dfrac{11}{7}$ D) $\dfrac{12}{11}$

3. $\dfrac{1}{13} + \dfrac{2}{13} + \dfrac{4}{13} = ?$

   A) $\dfrac{4}{13}$ B) $\dfrac{13}{7}$ C) $\dfrac{7}{13}$ D) $\dfrac{8}{13}$

4. $\dfrac{1}{7} + \dfrac{2}{7} + \dfrac{4}{7} = ?$

   A) $\dfrac{6}{7}$ B) $\dfrac{8}{7}$ C) $\dfrac{9}{7}$ D) 1

5. $\dfrac{7}{10} - \dfrac{3}{10} = ?$

   A) $\dfrac{21}{10}$ B) $\dfrac{10}{10}$ C) $\dfrac{2}{5}$ D) $\dfrac{3}{5}$

6. $\dfrac{17}{17} - \dfrac{7}{17} = ?$

   A) $\dfrac{10}{17}$ B) $\dfrac{24}{17}$ C) 1   D) $\dfrac{9}{17}$

7. $\dfrac{15}{29} - \dfrac{12}{29} = ?$

   A) $\dfrac{27}{29}$ B) $\dfrac{2}{29}$ C) $\dfrac{3}{29}$ D) $\dfrac{5}{29}$

8. $\left(\dfrac{3}{12} - \dfrac{1}{12}\right) + \left(\dfrac{4}{12} - \dfrac{1}{12}\right) = ?$

   A) $\dfrac{5}{12}$ B) $\dfrac{12}{5}$ C) $\dfrac{1}{12}$ D) $\dfrac{3}{12}$

9. $\left(\dfrac{5}{7} - \dfrac{1}{7}\right) + \left(\dfrac{4}{7} - \dfrac{2}{7}\right) = ?$

   A) $\dfrac{5}{7}$ B) $\dfrac{6}{7}$ C) $\dfrac{7}{7}$ D) $\dfrac{7}{6}$

10. $\left(\dfrac{9}{36} - \dfrac{1}{36}\right) + \left(\dfrac{8}{36} - \dfrac{2}{36}\right) = ?$

    A) $\dfrac{6}{36}$ B) $\dfrac{7}{18}$ C) $\dfrac{9}{36}$ D) $\dfrac{8}{36}$

FRACTION (ADDITION AND SUBTRACTION)
DIFFERENT DENOMINATOR

$$\frac{x}{a} + \frac{y}{b} = \frac{x \cdot b + a \cdot y}{a \cdot b}$$

Example:
$$\frac{7}{6} - \frac{3}{5} = \frac{7 \cdot 5 - 6 \cdot 3}{6 \cdot 5}$$
$$= \frac{35 - 18}{30} = \frac{17}{30}$$

Example:
$$\frac{2}{7} + \frac{1}{3} = \frac{3 \cdot 2 + 7 \cdot 1}{7 \cdot 3}$$
$$= \frac{6 + 7}{21} = \frac{13}{21}$$

TEST – 17

1. $\frac{7}{3} - \frac{1}{4} = ?$

   A) $\frac{23}{12}$ B) $\frac{31}{12}$ C) $\frac{25}{12}$ D) $\frac{24}{7}$

2. $\frac{7}{11} - \frac{3}{7} = ?$

   A) $\frac{77}{12}$ B) $\frac{15}{77}$ C) $\frac{72}{77}$ D) $\frac{16}{77}$

3. $\frac{1}{3} - \frac{1}{7} = ?$

   A) $\frac{4}{10}$ B) $\frac{4}{21}$ C) $\frac{5}{21}$ D) $\frac{1}{21}$

4. $\frac{4}{8} - \frac{1}{3} = ?$

   A) $\frac{1}{6}$ B) $\frac{1}{12}$ C) $\frac{1}{24}$ D) $\frac{3}{24}$

5. $\frac{7}{13} - \frac{1}{5} = ?$

   A) $\frac{32}{65}$ B) $\frac{36}{65}$ C) $\frac{34}{65}$ D) $\frac{22}{65}$

6. $\frac{1}{9} - \frac{1}{12} = ?$

   A) $\frac{1}{24}$ B) $\frac{1}{18}$ C) $\frac{1}{36}$ D) $\frac{5}{36}$

7. $\frac{3}{14} - \frac{3}{15} = ?$

   A) $\frac{1}{80}$ B) $\frac{1}{70}$ C) $\frac{1}{90}$ D) $\frac{3}{20}$

8. $\frac{5}{21} - \frac{2}{84} = ?$

   A) $\frac{5}{42}$ B) $\frac{7}{43}$ C) $\frac{8}{42}$ D) $\frac{9}{42}$

9. $\left(\frac{2}{3} - \frac{2}{5}\right)^2 = ?$

   A) $\frac{16}{225}$ B) $\frac{15}{125}$ C) $\frac{256}{225}$ D) $\frac{16}{25}$

10. $\left(\frac{1}{5} - \frac{1}{6}\right)^2 = ?$

   A) 900 B) $\frac{1}{30}$ C) $\frac{1}{900}$ D) $\frac{1}{800}$

FRACTION MULTIPLICATION

1) $\dfrac{a}{b} \times \dfrac{c}{d} = \dfrac{a \times c}{b \times d}$ Example: $\dfrac{5}{7} \times \dfrac{2}{3} = \dfrac{5 \times 2}{7 \times 3} = \dfrac{10}{21}$

2) $\dfrac{a}{b} \times \dfrac{c}{d} \times \dfrac{e}{f} = \dfrac{a \times c \times e}{b \times d \times f}$ Example: $\dfrac{2}{3} \times \dfrac{4}{5} \times \dfrac{7}{6} = \dfrac{2 \times 4 \times 7}{3 \times 5 \times 6} = \dfrac{56}{90}$

TEST – 18

1. $\dfrac{2}{3} \times \dfrac{4}{9} = ?$

   A) $\dfrac{6}{9}$ B) $\dfrac{6}{12}$ C) $\dfrac{6}{18}$ D) $\dfrac{8}{27}$

2. $\dfrac{7}{3} \times \dfrac{4}{11} = ?$

   A) $\dfrac{11}{33}$ B) $\dfrac{28}{11}$ C) $\dfrac{28}{33}$ D) $\dfrac{27}{33}$

3. $\dfrac{1}{3} \times \dfrac{2}{5} \times \dfrac{4}{7} = ?$

   A) $\dfrac{9}{105}$ B) $\dfrac{8}{105}$ C) $\dfrac{7}{110}$ D) $\dfrac{7}{105}$

4. $\dfrac{7}{11} \times \dfrac{11}{14} = ?$

   A) 1 B) $\dfrac{1}{2}$ C) 7 D) 49

5. $\dfrac{6}{22} \times \dfrac{11}{12} = ?$

   A) $\dfrac{1}{2}$ B) $\dfrac{1}{4}$ C) $\dfrac{1}{6}$ D) $\dfrac{1}{8}$

6. $\dfrac{11}{15} \times \dfrac{3}{33} = ?$

   A) $\dfrac{1}{5}$ B) $\dfrac{1}{10}$ C) $\dfrac{1}{15}$ D) $\dfrac{1}{30}$

7. $\left(\dfrac{2}{3}\right)^2 \times \left(\dfrac{3}{2}\right)^2 = ?$

   A) $\dfrac{6}{36}$ B) $\dfrac{36}{6}$ C) 1 D) $\dfrac{1}{4}$

8. $\dfrac{1}{1} \times \left(\dfrac{1}{2}\right)^2 \times \left(\dfrac{1}{3}\right)^2 = ?$

   A) $\dfrac{5}{6}$ B) $\dfrac{6}{5}$ C) $\dfrac{1}{18}$ D) $\dfrac{1}{36}$

9. $\dfrac{5}{1} \times \left(\dfrac{1}{5}\right)^2 \times \left(\dfrac{10}{2}\right)^2 = ?$

   A) 5 B) 10 C) 20 D) $\dfrac{1}{25}$

10. $\dfrac{11}{12} \times \dfrac{48}{33} \times \dfrac{3}{4} = ?$

    A) 1 B) 2 C) $\dfrac{1}{2}$ D) $\dfrac{2}{3}$

FRACTION (DIVISION)

$$\frac{a}{b} \div \frac{c}{d} = \frac{a}{b} \times \frac{d}{c}$$

Example: $\dfrac{5}{2} \div \dfrac{9}{7} = \dfrac{5}{2} \times \dfrac{7}{9} = \dfrac{35}{18}$

Example: $5 \div \dfrac{2}{3} = \dfrac{5}{1} \times \dfrac{3}{2} = \dfrac{15}{2}$

Example: $\dfrac{3}{5} \div 9 = \dfrac{3}{5} \div \dfrac{9}{1} = \dfrac{3}{5} \times \dfrac{1}{9} = \dfrac{3}{45}$

TEST – 19

1. $\dfrac{6}{5} \div \dfrac{2}{7} = ?$

   A) $\dfrac{42}{7}$ B) $\dfrac{42}{10}$ C) $\dfrac{43}{10}$ D) $\dfrac{12}{35}$

2. $\dfrac{11}{5} \div \dfrac{22}{10} = ?$

   A) 1   B) 2   C) $\dfrac{5}{11}$ D) $\dfrac{11}{5}$

3. $\dfrac{27}{5} \div \dfrac{9}{10} = ?$

   A) 4   B) $\dfrac{9}{5}$ C) $\dfrac{10}{9}$ D) 6

4. $\dfrac{1}{7} \div 7 = ?$

   A) 1   B) 49   C) $\dfrac{1}{49}$ D) 14

5. $6 \div \dfrac{1}{6} = ?$

   A) 1   B) $\dfrac{1}{12}$ C) $\dfrac{1}{36}$ D) 36

6. $\dfrac{11}{2} \div \dfrac{44}{5} = ?$

   A) $\dfrac{5}{8}$ B) $\dfrac{8}{5}$ C) $\dfrac{4}{7}$ D) $\dfrac{4}{11}$

7. $\dfrac{55}{6} \div \dfrac{44}{12} = ?$

   A) 2.5   B) 3   C) 3.5   D) 4

8. $(36 \div 4) \div (24 \div 4) = ?$

   A) 1   B) 1.5   C) 2   D) 3

9. $(12 \div 12 + 1) \div (14 \div 14 + 4) = ?$

   A) $\dfrac{5}{2}$ B) $\dfrac{5}{3}$ C) $\dfrac{2}{3}$ D) $\dfrac{2}{5}$

10. $(3 \div 3 + 3) \div (9 \div 9 + 9) = ?$

   A) $\dfrac{2}{5}$ B) $\dfrac{2}{7}$ C) $\dfrac{5}{2}$ D) $\dfrac{5}{7}$

FACTORING THE DIFFERENCE OF TWO SQUARES

$$a^2 - b^2 = (a + b) \times (a - b)$$

Example: $12^2 - 11^2 = (12 + 11) \times (12 - 11) = 23 \times 1 = 23$

Example: $99^2 - 90^2 = (99 + 90) \times (99 - 90) = 189 \times 9 = 1701$

Example: $100^2 - 80^2 = (100 - 80) \times (100 + 80) = 20 \times 180 = 3600$

TEST – 20

1. $15^2 - 14^2 = ?$
   A) 29　　B) 30　　C) 40　　D) 60

2. $30^2 - 20^2 = ?$
   A) 600　B) 500　　C) 400　　D) 200

3. $44^2 - 40^2 = ?$
   A) 326　B) 336　　C) 346　　D) 356

4. $21^2 - 20^2 = ?$
   A) 40　　B) 41　　C) 42　　D) 44

5. $19^2 - 9^2 = ?$
   A) 240　B) 260　　C) 270　　D) 280

6. $25^2 - 5^2 = ?$
   A) 700　　B) 600　　C) 500　　D) 400

7. $16^2 - 6^2 = ?$
   A) 190　B) 200　　C) 220　　D) 230

8. $33^2 - 3^2 = ?$
   A) 1080　B) 1090　　C) 1100　D) 1200

9. $\left(\dfrac{3}{4}\right)^2 - \left(\dfrac{1}{4}\right)^2 = ?$
   A) 1　　B) $\dfrac{1}{2}$　　C) $\dfrac{1}{3}$　　D) $\dfrac{1}{4}$

10. $18^2 - 8^2 = ?$
    A) 210　B) 230　　C) 260　　D) 280

DIVISION EQUATIONS

1) $\dfrac{x}{a} = b$, $x = a \cdot b$

Example: $\dfrac{x}{3} = 4$, $x = 3 \cdot 4 = 12$

Example: $\dfrac{x}{5} = 12$, $x = 5 \cdot 12 = 60$

2) $\dfrac{mx}{a} = 6$, $mx = a \cdot b$, $x = \dfrac{a \cdot b}{m}$

Example: $\dfrac{3x}{2} = 7$, $3x = 7 \cdot 2$, $3x = 14$, $x = \dfrac{14}{3}$

Example: $\dfrac{5x}{2} = 4$, $5x = 4 \cdot 3$, $5x = 12$, $x = \dfrac{12}{5}$

TEST – 21
(Solve the equation)

1. $\dfrac{x}{3} = 7$
A) 10 B) 21 C) 24 D) 4

2. $\dfrac{x}{9} = 12$
A) 108 B) 98 C) 78 D) 68

3. $\dfrac{n}{3} = 8$
A) 12 B) 24 C) 4 D) 2

4. $\dfrac{m}{3} = 13$
A) 39 B) 49 C) 40 D) 26

5. $\dfrac{y}{5} = 15$
A) 3 B) 15 C) 30 D) 75

6. $\dfrac{n}{7} = 27$
A) 189 B) 179 C) 169 D) 159

7. $\dfrac{a}{25} = 4$
A) 100 B) 110 C) 120 D) 140

8. $\dfrac{2m}{3} = 7$
A) 21 B) 42 C) $\dfrac{21}{2}$ D) $\dfrac{2}{21}$

9. $\dfrac{7m}{3} = 5$
A) $\dfrac{15}{8}$ B) $\dfrac{7}{15}$ C) $\dfrac{14}{7}$ D) $\dfrac{15}{7}$

10. $\dfrac{6m}{5} = 7$
A) $\dfrac{34}{6}$ B) $\dfrac{6}{34}$ C) $\dfrac{35}{7}$ D) $\dfrac{35}{6}$

ADDITION EQUATIONS

$x+a=b$, $x=b-a$

Example:
$$x+7=15$$
$$x=15-73$$
$$x=8$$

Example:
$$3x+4=25$$
$$x=25-4$$
$$3x=21$$
$$x=7$$

TEST – 22

(Solve the equation)

1. $x+6=17$
   A) 10   B) 11   C) 12   D) 17

6. $x + \dfrac{3}{2} = \dfrac{7}{2}$
   A) 1   B) 2   C) 3   D) $\dfrac{3}{4}$

2. $x+9=19$
   A) 9   B) 10   C) 11   D) 12

7. $x + \dfrac{11}{4} = \dfrac{16}{4}$
   A) $\dfrac{1}{5}$ B) $\dfrac{6}{4}$ C) $\dfrac{7}{4}$ D) $\dfrac{5}{4}$

3. $x+27=49$
   A) 20   B) 21   C) 22   D) 23

8. $x + \dfrac{7}{2} = \dfrac{22}{3}$
   A) $\dfrac{23}{6}$ B) $\dfrac{22}{7}$ C) $\dfrac{23}{5}$ D) $\dfrac{27}{5}$

4. $x+7=29$
   A) 20   B) 21   C) 24   D) 22

9. $n+3.6=9.6$
   A) 3   B) 3.1   C) 3.2   D) 6

5. $x+44=100$
   A) 52   B) 46   C) 56   D) 42

10. $x+7.2=14.4$
    A) 7   B) 7.1   C) 7.2   D) 7.3

MULTIPLYING MIXED NUMBERS

1) $\dfrac{a}{b} \times c = \dfrac{a}{b} \times \dfrac{c}{1} = \dfrac{a \times c}{b}$

Example: $\dfrac{3}{4} \times 24 = \dfrac{3}{4} \times \dfrac{24}{1} = \dfrac{24 \times 3}{4} = 18$

2) $\dfrac{a}{b} \times \left(m\dfrac{d}{f} \right) = \dfrac{a}{b} \times \left(\dfrac{f \cdot m + d}{f} \right) = \dfrac{a \cdot (f \cdot m + d)}{b \cdot f}$

Example: $\dfrac{5}{7} \times 6\dfrac{3}{4} = \dfrac{5}{7} \times \left(\dfrac{4 \times 6 + 3}{4} \right) = \dfrac{5}{7} \times \dfrac{27}{4} = \dfrac{135}{28}$

TEST – 23

1. $\dfrac{4}{7} \times 28 = ?$

   A) 16   B) 18   C) 20   D) 21

2. $\dfrac{7}{11} \times 44 = ?$

   A) 21   B) 28   C) 30   D) 36

3. $\dfrac{5}{6} \times 42 = ?$

   A) 30   B) 32   C) 34   D) 35

4. $72 \times \dfrac{8}{9} = ?$

   A) 54   B) 62   C) 63   D) 64

5. $\dfrac{11}{7} \times 35 = ?$

   A) 35   B) 45   C) 55   D) 65

6. $\dfrac{2}{3} \times 3\dfrac{1}{2} = ?$

   A) $\dfrac{14}{5}$ B) $\dfrac{14}{6}$ C) $\dfrac{5}{14}$ D) $\dfrac{17}{5}$

7. $7\dfrac{3}{2} \times \dfrac{4}{5} = ?$

   A) $\dfrac{88}{12}$ B) $\dfrac{85}{15}$ C) $\dfrac{34}{5}$ D) $\dfrac{97}{15}$

8. $25 \times 5\dfrac{3}{5} = ?$

   A) 120   B) 130   C) 135   D) 140

9. $28 \times 6\dfrac{3}{7} = ?$

   A) 160   B) 170   C) 175   D) 180

10. $36 \times 5\dfrac{16}{18} = ?$

    A) 162   B) 212   C) 182   D) 192

ADDING AND SUBTRACTING MIXED NUMBERS

$$a\frac{b}{c} + d\frac{e}{f} = \frac{c \times a + b}{c} + \frac{d \times f + e}{f}$$

Example: $4\frac{2}{3} + 7\frac{1}{2} = \frac{3 \times 4 + 2}{3} + \frac{2 \times 7 + 1}{2} = \frac{14}{3} + \frac{15}{2} = \frac{28 + 45}{6} = \frac{73}{6}$

TEST – 24

1. $2\frac{1}{3} + 3\frac{1}{2} = ?$

   A) $\frac{16}{3}$ B) $\frac{32}{7}$ C) $\frac{35}{6}$ D) $\frac{32}{5}$

2. $5\frac{3}{4} + 4\frac{2}{3} = ?$

   A) $\frac{126}{12}$ B) $\frac{125}{12}$ C) $\frac{127}{7}$ D) $\frac{143}{7}$

3. $2\frac{1}{3} + 4\frac{1}{7} = ?$

   A) $\frac{111}{21}$ B) $\frac{147}{8}$ C) $\frac{131}{7}$ D) $\frac{136}{21}$

4. $10\frac{1}{2} + 5\frac{2}{3} = ?$

   A) $\frac{97}{6}$ B) $\frac{98}{6}$ C) $\frac{97}{5}$ D) $\frac{96}{5}$

5. $12\frac{1}{5} + 13\frac{1}{4} = ?$

   A) $\frac{201}{20}$ B) $\frac{301}{20}$ C) $\frac{509}{20}$ D) $\frac{501}{20}$

6. $4\frac{2}{3} - 2\frac{1}{3} = ?$

   A) $\frac{7}{4}$ B) $\frac{7}{6}$ C) $\frac{7}{3}$ D) $\frac{3}{7}$

7. $9\frac{2}{3} - 6\frac{1}{2} = ?$

   A) $\frac{19}{6}$ B) $\frac{19}{5}$ C) $\frac{18}{12}$ D) $\frac{17}{6}$

8. $15\frac{2}{3} - 12\frac{1}{4} = ?$

   A) $\frac{1}{3}$ B) $41/12$ C) $\frac{1}{6}$ D) $\frac{1}{12}$

9. $17\frac{2}{5} - 16\frac{1}{4} = ?$

   A) $\frac{23}{20}$ B) $\frac{103}{20}$ C) $\frac{203}{20}$ D) $\frac{103}{5}$

10. $6\frac{1}{5} - 4\frac{1}{4} = ?$

    A) $\frac{5}{11}$ B) $\frac{7}{11}$ C) $\frac{39}{20}$ D) $\frac{11}{5}$

SUBTRACTION EQUATIONS

1) x-a=b, x=a+b

Example:
x-2.2=6.4
x=2.2+6.4
x=8.6

Example:
x-4=8
x=4+8=12

2) $mx - a = b$

$mx = a + b$

$x = \dfrac{a + b}{m}$

Example:
2x-7=3
2x=7+3
2x=10
$x = \dfrac{10}{2} = 5$

Example:
3x-3=21
3x=21+3
3x=24
$x = \dfrac{24}{3} = 8$

TEST – 25
(Solve the equation)

1. x-5=12
   A) 7
   B) 8
   C) 14
   D) 17

2. x-8=20
   A) 12
   B) 16
   C) 22
   D) 28

3. x-9=19
   A) 20
   B) 18
   C) 28
   D) 30

4. x-3.7=6.3
   A) 10
   B) 9
   C) 8
   D) 7.7

5. x-6.25=4.75
   A) 11
   B) 10
   C) 10.25
   D) 9.75

6. 3x-3=30
   A) 8
   B) 9
   C) 10
   D) 11

7. 4x+4=14
   A) 2
   B) 2.3
   C) 2.4
   D) 2.5

8. 5x-5=75
   A) 12
   B) 14
   C) 15
   D) 16

9. $3x - \dfrac{4}{11} = \dfrac{18}{11}$
   A) $\dfrac{3}{2}$
   B) $\dfrac{2}{3}$
   C) $\dfrac{1}{3}$
   D) $\dfrac{2}{5}$

10. $7x - \dfrac{7}{12} = \dfrac{5}{12}$
   A) 1
   B) $\dfrac{1}{3}$
   C) 7
   D) $\dfrac{1}{7}$

MULTIPLICATION EQUATION

$ax=b, \quad x = \dfrac{b}{a}$

Example: $6x=28, \quad x = \dfrac{28}{6} = \dfrac{14}{3}$

Example: $7x=21, \quad x = \dfrac{21}{7} = 3$

Example: $0.3a=1.2 \quad a = \dfrac{1.2}{0.3} = \dfrac{12}{3} = 4$

Example: $0.4a=16 \quad a = \dfrac{16}{0.4} = \dfrac{160}{4} = 40$

TEST – 26

(Solve the equation)

1. 6m=48
   A) 6   B) 7   C) 8   D) 9

2. 8m=32
   A) 2   B) 3   C) 4   D) 6

3. 6m=144
   A) 18   B) 19   C) 20   D) 24

4. 8a=96
   A) 10   B) 11   C) 12   D) 9

5. 10n=180
   A) 18   B) 10   C) 11   D) 12

6. 5m=120
   A) 20   B) 21   C) 22   D) 24

7. 2.7m=54
   A) 14   B) 15   C) 16   D) 20

8. 1.2m=6
   A) 5   B) 6   C) 7   D) 8

9. 0.4m=2.8
   A) 2   B) 3   C) 4   D) 7

10. 0.25n=20
    A) 60   B) 70   C) 80   D) 90

PERCENT

Write fraction as a percent.

Example: $\dfrac{3}{4} = \dfrac{3}{4} \times 100\% = \dfrac{3 \times 100}{4}\% = \dfrac{300}{4}\% = 75\%$

- $\dfrac{a}{b}$ fraction convert to % Percent

$\dfrac{a}{b} \times 100\% = \dfrac{a \times 100}{b}\%$

Example: $\dfrac{8}{25}$ fraction convert to percent.

$\dfrac{8}{25} = \dfrac{8}{25} \times 100\% = \dfrac{8 \times 100}{25}\% = \dfrac{800}{25}\% = 32\%$

TEST – 27
(Write each fraction as a percent)

1. $\dfrac{13}{50} = \dots\%$
   A) 13%   B) 26%   C) 30%   D) 42%

2. $\dfrac{7}{50} = \dots\%$
   A) 7%   B) 21%   C) 14%   D) 28%

3. $\dfrac{6}{4} = \dots\%$
   A) 60%   B) 80%
   C) 120%   D) 150%

4. $\dfrac{7}{2} = \dots\%$
   A) 70%   B) 120%
   C) 140%   D) 350%

5. $\dfrac{23}{25} = \dots\%$
   A) 92%   B) 82%   C) 72%   D) 60%

6. $2\dfrac{1}{4} = \dots\%$
   A) 200%   B) 225%
   C) 230%   D) 280%

7. $\dfrac{62}{100} = \dots\%$
   A) 62%   B) 64%   C) 32%   D) 31%

8. $\dfrac{4}{5} = \dots\%$
   A) 60%   B) 70%   C) 80%   D) 90%

9. $\dfrac{6}{200} = \dots\%$
   A) 6%   B) 3%   C) 12%   D) 90%

10. $\dfrac{8}{400} = \dots\%$
    A) 2%   B) 4%   C) 6%   D) 8%

PERCENT OF A NUMBER

$$a\% \text{ of } b = \frac{a}{100} \times b$$

- Find the 6% of 20 $\qquad$ $6\% \times 20 = \frac{6}{100} \times 20 = \frac{120}{100} = 1.2$

- Find the 15% of 200 $\qquad$ $15\% \times 200 = \frac{15}{100} \times 200 = 30$

TEST – 28

(Find the percent of each number)

1. 7% of 20:
   A) 10 $\qquad$ B) 14 $\qquad$ C) 28 $\qquad$ D) 1.4

2. 4% of 40:
   A) 30 $\qquad$ B) 160 $\qquad$ C) 1.6 $\qquad$ D) 20

3. 30% of 30:
   A) 10 $\qquad$ B) 30 $\qquad$ C) 90 $\qquad$ D) 9

4. 120% of 40:
   A) 48 $\qquad$ B) 60 $\qquad$ C) 80 $\qquad$ D) 96

5. 5% of 5:
   A) 25 $\qquad$ B) 50 $\qquad$ C) 5 $\qquad$ D) 0.25

6. What number is 6% of 60?
   A) 3.6 $\qquad$ B) 100 $\qquad$ C) 1000 $\qquad$ D) 800

7. What number is 20% of 22?
   A) 4.4 $\qquad$ B) 96 $\qquad$ C) 100 $\qquad$ D) 110

8. What number is 28% of 56?
   A) 15.68 $\quad$ B) 80 $\qquad$ C) 100 $\qquad$ D) 200

9. What number is 22% of 440?
   A) 500 $\qquad\qquad$ B) 800
   C) 96.8 $\qquad\qquad$ D) 2000

10. What number is 9% of 18?
    A) 200 $\qquad$ B) 180 $\qquad$ C) 150 $\qquad$ D) 1.62

FINDING THE PERCENT ONE NUMBER IS OF ANOTHER

TEST – 29

1. What percent of 36 is 9?
    A) 25%   B) 30%   C) 40%   D) 45%

2. What percent of 25 is 5?
    A) 10%   B) 20%   C) 25%   D) 30%

3. 16 is what percent of 50?
    A) 16%   B) 20%   C) 32%   D) 40%

4. 7 is what percent of 25?
    A) 7%   B) 14%   C) 21%   D) 28%

5. 14 is what percent of 70?
    A) 10%   B) 20%   C) 30   D) 40%

6. What percent of 50 is 24?
    A) 48%   B) 40%   C) 50%   D) 60%

7. What percent of 25 is 4?
    A) 4%   B) 8%   C) 12%   D) 16%

8. 40 is what percent of 160?
    A) 25%   B) 30%   C) 40%   D) 50%

9. 12 is what percent of 60?
    A) 20%   B) 25%   C) 30%   D) 40%

10. What percent of 65 is 13?
    A) 5%   B) 10%   C) 15%   D) 20%

INTEGER AS EXPONENTS

$10^0 = 1$ $10^1 = 10$ $10^2 = 10 \times 10 = 100$ $10^3 = 10 \times 10 \times 10 = 1000$

$10^{-1} = \dfrac{1}{10}$ $10^{-2} = \dfrac{1}{100}$ $10^{-3} = \dfrac{1}{1000}$

$10^{-4} = \dfrac{1}{10000}$

$a^0 = 1$ $a^1 = a$ $a^2 = a \times a$ $a^3 = a \times a \times a$

$a^{-1} = \dfrac{1}{a}$ $a^{-2} = \dfrac{1}{a^2}$ $a^{-3} = \dfrac{1}{a^3}$

TEST – 30

1. $3 \times \left(3^{-1}\right) = ?$
   A) 6   B) 0   C) 1   D) 9

2. $1000 \times 10^{-3} = ?$
   A) 1   B) 0   C) 100   D) 10^6

3. $100 + 10^{-2} = ?$
   A) 1   B) 200   C) 100.01   D) 0

4. $\dfrac{1}{10^4} = ?$
   A) 10^{-4} B) 1000   C) 10000   D) 400

5. $\dfrac{1}{3^7} = ?$
   A) 3^7 B) 3^{-7} C) -21   D) 7^{-3}

6. $10^{-11} = ?$
   A) $\dfrac{1}{10^{11}}$ B) $\dfrac{1}{11^{10}}$ C) 10^{11} D) -110

7. $3^{14} \times 3^{-14} = ?$
   A) 1   B) 2   C) 3^{28} D) 3^{14}

8. $49 \times 7^{-2} = ?$
   A) 1   B) 2   C) 7   D) 49

9. $3^0 + 3^1 + 3^2 + 3^3 = ?$
   A) 36   B) 38   C) 39   D) 40

10. $\left(2^{-1} + 2^1\right) = ?$
    A) 0   B) 1   C) $\dfrac{4}{25}$ D) 2.5

SQUARE ROOTS

$$a^2 = a \times a \qquad 15^2 = 15 \times 15 = 225$$
$$6^2 = 6 \times 6 = 36 \qquad 11^2 = 11 \times 11 = 121$$

$$\sqrt{4} = 2, \quad \sqrt{9} = 3, \quad \sqrt{16} = 4, \quad \sqrt{25} = 5$$
$$\sqrt{36} = 6, \quad \sqrt{49} = 7, \quad \sqrt{64} = 8, \quad \sqrt{81} = 9$$
$$\sqrt{100} = 10, \quad \sqrt{121} = 11, \quad \sqrt{144} = 12$$

TEST – 31

1. $5^2 + 4^2 + 3^2 + 2^2 + 1^2 = ?$
   A) 56   B) 55   C) 54   D) 53

2. $17^2 + 16^2 = ?$
   A) 545   B) 544   C) 543   D) 530

3. $\left(\dfrac{2}{7}\right)^2 = ?$
   A) $\dfrac{4}{14}$ B) $\dfrac{4}{28}$ C) $\dfrac{49}{4}$ D) $\dfrac{4}{49}$

4. $\left(\dfrac{3}{5}\right)^2 = ?$
   A) $\dfrac{6}{10}$ B) $\dfrac{9}{15}$ C) $\dfrac{9}{25}$ D) $\dfrac{6}{25}$

5. $\sqrt{144} + \sqrt{121} + \sqrt{289} = ?$
   A) 33   B) 36   C) 38   D) 40

6. $\sqrt{81} + \sqrt{225} = ?$
   A) 21   B) 22   C) 24   D) 26

7. $\sqrt{0.25} = ?$
   A) $\dfrac{1}{2}$ B) $\dfrac{1}{3}$ C) $\dfrac{1}{4}$ D) $\dfrac{1}{5}$

8. $\sqrt{0.04} + \sqrt{0.09} = ?$
   A) 0.7   B) 0.6   C) 0.5   D) 0.4

9. $\sqrt{0.81} \div \sqrt{0.09} = ?$
   A) 1   B) 2   C) 3   D) 4

10. $\sqrt{0.36} \div \sqrt{0.49} = ?$
    A) $\dfrac{7}{6}$ B) $\dfrac{6}{7}$ C) $\dfrac{3}{7}$ D) $\dfrac{7}{3}$

ORDER FO OPERATION

P – parentheses
E – exponents
M – multiply
D – divide
A – add
S – subtract

Simplify: $20 + 4^2 + (6-4) \times 4$.

$$20 + 4^2 + (6-4) \times 4 = 20 + 16 + 2 \times 7 = 20 + 16 + 14 = 50$$

TEST – 32

1. $4 + 24 \div 4 = ?$
   A) 7   B) 8   C) 9   D) 10

2. $36 \div 6 + 6 = ?$
   A) 7   B) 8   C) 9   D) 12

3. $(3^2 + 9) \times 4^2 = ?$
   A) 288   B) 268   C) 248   D) 238

4. $(28 \div 7 + 7)^2 = ?$
   A) 4   B) 8   C) 16   D) 121

5. $(16 \div 4^2)^2 + (36 \div 6 - 4)^2 = ?$
   A) 8   B) 7   C) 6   D) 5

6. $24 + 24 \div 24 + 36 - 36 \div 12 = ?$
   A) 38   B) 48   C) 58   D) 68

7. $(4^2 \div 4)^2 + (27 \div 9 + 3)^2 = ?$
   A) 32   B) 42   C) 52   D) 62

8. $(64 \div 8^2) + (144 \div 12) \times 5^2 = ?$
   A) 301   B) 290   C) 281   D) 271

9. $(81 \div 27 + 3) \times (16 \div 4 + 4) = ?$
   A) 48   B) 42   C) 36   D) 32

10. $(12 \div 2 + 2) \times (15 \div 5 + 5) = ?$
    A) 60   B) 64   C) 68   D) 72

DISTANCE, SPEED AND TIME

Distance = d, Speed = s,   Time = t

$$\boxed{d = s \times t} \qquad s = \frac{d}{t} \qquad t = \frac{d}{s}$$

TEST – 33

1. d=50-mile, t=2 h, s=?
    A) 20   B) 25   C) 30   D) 32

2. d=240-mile, s=40 m/h, t=?
    A) 6   B) 8   C) 9   D) 12

3. s=30 m/h, t=8 h, d=?
    A) 7.5-mile   B) 120 mile
    C) 240-mile   D) 260 mile

4. d=300 mile, t=5 h, s=?
    A) 60 m/h   B) 70 m/h
    C) 80 m/h   D) 90 m/h

5. Truck speed is 32 miles per hour, and the track for 5 hours. How far will the truck travel?
    A) 140   B) 150   C) 160   D) 170

6. Luis traveled 320 km at an average speed at 40 km/h. Find the time.
    A) 9   B) 8   C) 7   D) 6

7. Ronaldo drove his car at 70 miles per hour for 5 hours. How far did he travel?
    A) 320   B) 350   C) 360   D) 380

8. Anderson can drive his bicycle at 8 miles per hour. At this rate, how long will it take him to ride 48 miles?
    A) 6   B) 7   C) 8   D) 9

9. Javier traveled 480 km at an average speed at 60 km/h. Find the time.
    A) 6   B) 7   C) 8   D) 9

10. Susana drove his car at 80 miles per hour for 7 hours. How far did he travel?
    A) 480   B) 490   C) 520   D) 560

WRITE AN EQUATION

1) Six times a number equal 90.   *Solution:* 6×a=90, 6a=90

2) A number divided by -4 equal 12.   *Solution:* $\dfrac{a}{-4}=12$

3) 20 more than a number equal 24.   *Solution:* a+20=24

TEST – 34
(Write as an equation)

1. Seven times a number equal 42.
   A) 7a = 42   B) 7 ÷ a = 42
   C) 7 + a = 42   D) a ÷ 7 = 42

2. Nine times a number equals 108.
   A) 9 + a = 108   B) 9a = 108
   C) 9 − a = 108   D) $\dfrac{9}{a}=108$

3. A number divided by 4 equals 24.
   A) 4a = 24   B) $\dfrac{a}{4}=24$
   C) a ÷ 4 = 24   D) 4 − a = 24

4. A number less than 9 equal 23.
   A) a + 9 = 23   B) a − 9 = 23
   C) 9 − a = 23   D) a − 23 = 9

Write each phrase as an algebraic expression

5. m plus 9.
   A) 9m   B) 9+m   C) 9-m   D) m-9

6. The sum of 1 and m.
   A) m   B) m+1   C) m-1   D) 1-m

7. y less than 22.
   A) y-22   B) 22-y   C) 22y   D) y ÷ 22

8. m multiply by square of 3.
   A) 3m   B) 3+m   C) 9+m   D) 9m

9. The quotient of m and 4.
   A) 4m   B) 4-m   C) $\dfrac{4}{m}$ D) $\dfrac{m}{4}$

10. The product of 12 and m.
    A) 12+m   B) 12-m
    C) 12m   D) m-12

EXPLORE MULTIPLICTION
DIVISION BY DECIMALS

TEST – 35
(Find each product or quotient)

1. $0.3 \times 0.7 = ?$
   A) 21   B) 2.1   C) 0.21   D) 0.021

6. $0.24 \div 0.6 = ?$
   A) 0.2   B) 0.3   C) 0.4   D) 4

2. $0.6 \times 0.3 = ?$
   A) 12   B) 1.8   C) 0.18   D) 0.09

7. $0.15 \div 0.5 = ?$
   A) 3   B) 0.3   C) 5   D) 0.5

3. $0.9 \times 0.5 = ?$
   A) 0.15   B) 0.45   C) 0.54   D) 0.13

8. $2.5 \div 0.5 = ?$
   A) 1   B) 10   C) 15   D) 5

4. $0.12 \times 0.21 = ?$
   A) 0.0252   B) 0.252
   C) 2.52   D) 25.2

9. $0.36 \div 0.12 = ?$
   A) 12   B) 10   C) 3   D) 8

5. $0.8 \times 0.7 = ?$
   A) 0.54   B) 0.56   C) 5.6   D) 6.5

10. $0.72 \div 0.09 = ?$
    A) 12   B) 10   C) 9   D) 8

MULTIPLY AND DIVIDE BY POWERS OF TEN

$4.2 \times 10^1 = 42$

$3.123 \times 10^3 = 3.123 \times 1000 = 3123$

$24.36 \div 10^2 = 0.2436$

$6.25 \times 10^2 = 6.25 \times 100 = 625$

$6.3 \div 10^1 = 0.63$

$142.3 \div 10^3 = 0.1423$

TEST – 36
(Find each product or quotient)

1. $3.6 \times 10^2 = ?$
   A) 36   B) 360   C) 3600   D) 0.36

   A) 3624   B) 0.3624
   C) 0.03624   D) 36.24

2. $0.48 \times 10^0 = ?$
   A) 48   B) 4.8   C) 0.48   D) 0.048

7. $2.7 \div 10^{-1} = ?$
   A) 27   B) 0.27   C) 270   D) 0.027

8. $0.45 \times 10^{-2} = ?$
   A) 0.45   B) 0.045
   C) 0.0045   D) 45

3. $0.321 \times 10^3 = ?$
   A) 321   B) 3210   C) 32.1   D) 3.21

9. $3.9 \times 10^{-1} \times 10^0 = ?$
   A) 1   B) 39   C) 3.9   D) 0.39

4. $13.24 \times 10^4 = ?$
   A) 132.4   B) 1324
   C) 13240   D) 132400

10. $2.4 \times 10^{-2} \times 10^2 = ?$
    A) 2.4   B) 24   C) 0.24   D) 0.024

5. $0.0364 \times 10^2 = ?$
   A) 0.364   B) 364   C) 36.4   D) 3.64

6. $36.24 \div 10^{-2} = ?$

FIND AN EQUATION

TEST – 37

1. The sum of 6 and a number.

   A) 6n   B) 6+n   C) $\dfrac{6}{n}$ D) $\dfrac{n}{6}$

2. Three times the number of books plus four.
   A) 3b+3   B) 3b+4   C) 3b-4   D) 4b+3

3. 7 less than three times a number.
   A) 7+3n   B) 7-3n   C) 3n-7   D) 3n+7

4. Jack has 25 pens. His brother has m times more pens than this. Write an expression for how many pens his brother has.

   A) 25   B) $\dfrac{25}{m}$ C) 25m   D) 25-m

5. Jack has 50 SAT math books. If he sells m math book per day for 6 days, how many math books will he left? Choose an expression.
   A) 50m   B) 50+6m
   C) 50-6m   D) 6m-50

6. Six times a number plus the sum of three times a number and -4.
   A) $6(2x-4)$ B) $6x+(3x-4)$
   C) $4x(2x-6)$ D) $4(6x-3)$

7. 8 more than twice a number.
   A) 8x+2   B) 2x+3
   C) 2x-8   D) 2x+8

8. 6 decreased by the quotient of a number and 8.

   A) $6+\left(\dfrac{n}{8}\right)$ B) $6+8n$

   C) $6-\left(\dfrac{n}{8}\right)$ D) $8-\left(\dfrac{n}{6}\right)$

9. 18 increased by a number is 30.
   A) 18n=30   B) 18-n=30
   C) 18n+30=0   D) 18+n=30

10. Two times the sum of m and 6.
    A) 2(m+6)   B) 6(m+2)
    C) 2(m-6)   D) 6(m-6)

PERCENT

TEST – 38

1. Find 20% of 24.
   A) 4   B) 4.2   C) 4.8   D) 4.9

2. Find $15\frac{1}{2}\%$ of 300.

   A) 46   B) 46.5   C) 47   D) 48

3. 8 is 12% of what number?
   A) 60   B) 62   C) 66   D) 66.6

4. 120 is what percent of 1600?
   A) 7.5   B) 6.5   C) 0.75   D) 0.075

5. 28 is what percent of 140?
   A) 15%   B) 20%   C) 30%   D) 35%

6. James has a job working 10 hours per day. If he only worked 35% of his workday, how many minutes did he work?
   A) 150 min   B) 180 min
   C) 210 min   D) 240 min

7. Write 0.8% as a decimal.
   A) 8   B) 0.8   C) 0.08   D) 0.008

8. The SAT book`s original price is $60. It is sold for $45. What is the rate of discount?
   A) 15%   B) 20%   C) 25%   D) 30%

9. How much interest is earned in 3 years on a investment of $30000 at 10%?
   A) $900   B) $9000
   C) $12000   D) $15000

10. A book`s original price is $40. Markdown is 20%. Find the sale price.
    A) $48   B) $42   C) $32   D) $28

WORD PROBLEM – CONSECUTIVE INTEGER PROBLEMS

TEST – 39

1. The sum of two consecutive integers is 33. Find the product of these integers.
   A) 272   B) 282   C) 262   D) 242

2. The sum of three consecutive integers is 36. Find the biggest number.
   A) 11   B) 12   C) 13   D) 14

3. The sum of three consecutive integers is 6m. find the second number.
   A) m   B) 2m
   C) 2m+2   D) 2m+4

4. The sum of two odd consecutive integers is 64. Find the product of these numbers.
   A) 1021   B) 1022
   C) 1023   D) 1024

5. The sum of two numbers 66 and their difference is 6. What is the value of the smaller number?
   A) 26   B) 29   C) 30   D) 36

6. The positive difference between the squares of two consecutive odd integers is 32. Find the big number.
   A) 9   B) 8   C) 7   D) 6

7. How many odd integers are between 46 and 136?
   A) 45   B) 46   C) 47   D) 48

8. The sum of three consecutive integers is 30. Find the biggest number.
   A) 10   B) 12   C) 13   D) 11

9. If twice the smaller of two consecutive integers is added to the large, the result is 91. Find the product of the numbers.
   A) 960   B) 950   C) 940   D) 930

10. The square of the sum of two integers is 225 and these numbers` ratio is 3:4. Find the biggest number.
    A) 9   B) 11   C) 12   D) 15

WORD PROBLEM – FRACTION

TEST – 40

1. Find the ratio of $\frac{1}{5}$ to $\frac{4}{5}$.

   A) $\frac{1}{4}$ B) $\frac{1}{25}$ C) $\frac{4}{25}$ D) 4

2. If $\frac{8}{11}$ of a number is 120, what is the number?
   A) 145   B) 155   C) 165   D) 175

3. 27 is $\frac{3}{5}$ of what number?

   A) 35   B) 39   C) 40   D) 45

4. If a MATHCOUNTS club consists of 9 girls and 14 boys, what part of the club is girls?

   A) $\frac{9}{14}$ B) $\frac{14}{9}$ C) $\frac{9}{23}$ D) $\frac{14}{23}$

5. Simplify the expression:
   $$\left(-\frac{3}{21}\right) \cdot \left(\frac{56}{12}\right)$$

   A) $\frac{2}{3}$ B) $\frac{3}{2}$ C) $-\frac{2}{3}$ D) $-\frac{3}{2}$

6. Luis paid $64.32 for 8 books. How much did he pay per book?
   A) $8.40   B) $8.04
   C) $16.04   D) $8.24

7. The four sides of a garden measure $4\frac{2}{3}$ meters, $8\frac{1}{3}$ meters, $12\frac{1}{3}$ meters and $10\frac{2}{3}$ meters. Find the length of fence needed to enclose the garden.

   A) 26   B) 36   C) $26\frac{2}{3}$ D) $36\frac{1}{3}$

8. Javier workbook has 300 questions. Ronaldo solved $\frac{1}{6}$ of questions on Monday and $\frac{1}{4}$ of questions on Tuesday. How many questions did Jack solve?
   A) 100   B) 115   C) 120   D) 125

9. The new geometry book was on sale for $\frac{4}{7}$ of the original price. If the original price was $77 what was the sale price?
   A) $33   B) $44   C) $55   D) $66

10. Which is rational number but not an integer?

    A) -9   B) 2   C) $\frac{\sqrt{27}}{\sqrt{3}}$ D) $\frac{\sqrt{3}}{\sqrt{27}}$

WORD PROBLEM – PERCENT INCREASE AND DECREASE

TEST – 41

1. The price of books increases from $40 to $50. What are the percent increases?
    A) 20%   B) 25%   C) 30%   D) 40%

2. A square's perimeter is 40 cm. if the perimeter increases 20%, find the new square's area.
    A) 122   B) 144   C) 16   D) 289

3. What is the percent increase of the area?
    A) 22%   B) 30%   C) 40%   D) 44%

4. The number of students in math league club decreased from 24 to 16. What is the percent decrease?
    A) 11%   B) 22%   C) 33%   D) 44%

5. The number of books published decreased from 30000 to 24000. What is the percent decrease?
    A) 10%   B) 20%   C) 25%   D) 30%

6. A city population went from 42000 to 48000 in 6 years. What was the percent of change?
    A) 12% increase   B) 12% decrease
    C) 14% increase   D) 14% decrease

7. What percent of the figure is shaded?

    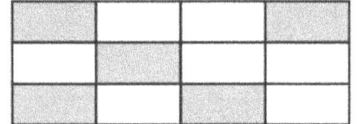

    A) 36%   B) 38%   C) 40%   D) 42%

8. If a square's side increases 10%, what is the percent of change of the area?
    A) 10%   B) 16%   C) 20%   D) 21%

9. A book marked at $60 is sold for $45. What is the percent of the markdown?
    A) 20%   B) 25%   C) 30%   D) 35%

10. A book's price is $80. The seller made two time a 10% discount. Find the final price.
    A) $64.8   B) $66.8
    C) $62.4   D) $61.2

DIRECT AND INVERSE VARIATION

TEST – 42

1. If y varies directly as x and y=15 when x=24, then what is the y when x=96?
   A) 30   B) 36   C) 60   D) 45

2. If y varies inversely as x and y=16 when x=6, then what is y when x=12?
   A) 6   B) 8   C) 9   D) 32

3. Which equation represents a direct variation?
   A) y=x+2   B) y=4x
   C) y=6-x   D) $y = \dfrac{4}{x}$

4. y varies inversely with the square of x. If y=9 when x=5, find y when x=4.
   A) $\dfrac{225}{8}$ B) $\dfrac{225}{4}$ C) $\dfrac{225}{16}$ D) $\dfrac{125}{8}$

5. ABC is right triangle. $\angle A$ is a right angle. The ratio of $\angle B$ to $\angle C$ is 2 to 3. Find $\angle B$.
   A) 26   B) 30   C) 36   D) 45

6. 24 workers can complete a new home construction in 14 days. How many workers will be required to complete this task in 21 days?
   A) 16   B) 18   C) 20   D) 21

7. 9 workers can paint a schools` walls in 12 days. How many workers will be required to complete this school wall paint in 6 days?
   A) 12   B) 14   C) 16   D) 18

8. y is directly proportional to x when x-45, y=60. Find y when x=60.
   A) 80   B) 70   C) 75   D) 90

9. y is proportional to the cube of x. y=40 when x=2. Find the value of x when y=625.
   A) 5   B) 10   C) 15   D) 25

10. 18 workers will take 900 hours to finish a small new house. How long will take 6 workers to finish the new house?
    A) 2400   B) 2600
    C) 2700   D) 3000

WORD PROBLEMS – AGE PROBLEMS

TEST – 43

Jack is now 16 years old and Mario is now 10 years old.

	Jack	Mario	Sum of ages
Now	16	10	26
After 5 years	M	N	K
4 years ago	X	Y	Z

1. Find the K.
   A) 36   B) 35   C) 34   D) 33

2. Find the M.
   A) 21   B) 20   C) 19   D) 18

3. N+K=?
   A) 39   B) 38   C) 51   D) 36

4. X+Y+Z=?
   A) 39   B) 38   C) 36   D) 35

5. The ratio of Jack`s to Mario`s age is 1:3. If the sum of their ages is 24, what is Mario`s age?
   A) 6   B) 7   C) 12   D) 18

6. Jack is 24 years old and Mario is 18 years old. Calculate the ratio of Jack`s age to Mario`s age 8 years from now.
   A) $\frac{16}{15}$ B) $\frac{15}{16}$ C) $\frac{16}{13}$ D) $\frac{13}{16}$

7. The sum of the ages of 4 siblings is 86. Find the sum of their ages in 5 years.
   A) 106   B) 107   C) 108   D) 110

8. A father is 36 years old, and his daughter is 14 years old. In how many years will the ratio of their ages be 2:1?
   A) 4   B) 5   C) 6   D) 8

9. A 64 years old father has three children whose ages add to 16. In how many years will the father`s age be 3 times the sum of his children`s ages?
   A) 1   B) 2   C) 3   D) 4

10. Three brothers have ages in ratio 1:2:3. If the sum of the brothers` ages is 36, find the eldest brother`s age.
    A) 10   B) 12   C) 16   D) 18

WORD PROBLEMS – RATE PROBLEMS

TEST – 44

1. Which of the following ration is not equivalent to the other three?

   A) $\dfrac{7}{9}$ 　　　　 B) 7 to 9

   C) 14 to 18 　　 D) $\dfrac{21}{26}$

2. Susana bought 5 kilograms walnuts for $8.50. What is the unit price per kilogram?

   A) $1.7/kg 　　　 B) $17/kg
   C) $3.4/kg 　　　 D) $1.8/kg

3. A machine washes 3 carpets in 45 minutes. Which expression equals the unit rate per hour?

   A) 1 　　 B) 2 　　 C) 3 　　 D) 4

4. Ronaldo bought 6 SAT math books for $18.90. what is the unit price?

   A) $3 　 B) $3.15 　 C) $2 　 D) $2.8

5. Luis bought 3 dozen packages of pencils for $3.60. What is the unit price per pencil?

   A) $0.6 　 B) $0.2 　 C) $0.25 　 D) $0.10

6. Jack changed 24 tires in 15 minutes. Find how many tires per hour he can change.

   A) 24 tires/1 hour
   B) 96 tires/1 hour
   C) 36 tires/1 hour
   D) 42 tires/1 hour

7. If 8.4 pounds of apples cost $4.20, what is the unit price?

   A) $0.50 　　　 B) $0.60
   C) $0.70 　　　 D) $0.80

8. If Javier drives 220 miles in 4 hours, how far can he drive in 6 hours?

   A) 330 miles 　　　 B) 300 miles
   C) 2890 miles 　　 D) 270 miles

Number of cars and washes

Cars	4	6	8
Washes (min time)	60	90	X

The table shows the number of the cars and wash times.

9. Find the X.

   A) 100 　　 B) 110 　　 C) 120 　　 D) 140

10. Which of the following represents the wash times?

   A) cars+15 　　　 B) cars-15
   C) cars×15 　　　 D) cars×30

TEST – 45

1. 48+96=?
   A) 153   B) 155   C) 143   D) 144

2. 99+88+77=?
   A) 260   B) 261   C) 264   D) 263

3. 77+44+33=?
   A) 154   B) 164   C) 174   D) 184

4. 88+87+86=?
   A) 231   B) 241
   C) 251   D) 261

5. 984+177+213=?
   A) 1260   B) 1360   C) 1374   D) 1384

6. 124+874+777=?
   A) 1775   B) 1675   C) 1572   D) 1677

7. 999-777+222=?
   A) 333   B) 222   C) 444   D) 555

8. 334+433+127=?
   A) 854   B) 864   C) 894   D) 874

9. 1984+692+813=?
   A) 3489   B) 3479   C) 3579   D) 3679

10. 8417+9627+1217=?
    A) 19261   B) 18361
    C) 17361   D) 16371

TEST – 46

1. 827-644=?
   A) 183   B) 184   C) 18   D) 196

2. 784-244=?
   A) 340   B) 440   C) 448   D) 540

3. 962-374=?
   A) 588   B) 377   C) 488   D) 388

4. 623-443=?
   A) 160   B) 170   C) 180   D) 190

5. 927-216=?
   A) 612   B) 611   C) 711   D) 811

6. 9782-2121=?
   A) 7662   B) 7912   C) 8622   D) 7661

7. 6412-3147=?
   A) 3265   B) 3165   C) 3642   D) 3427

8. 8743-1473
   A) 7270   B) 7327   C) 7643   D) 6270

9. 5543-3291=?
   A) 2253   B) 2252   C) 2251   D) 3252

10. 1672-1247=?
    A) 421   B) 422   C) 423   D) 425

TEST – 47

1. 74×8=?
   A) 492   B) 522   C) 624   D) 592

2. 99×9=?
   A) 791   B) 781   C) 891   D) 991

3. 84×8=?
   A) 671   B) 672   C) 682   D) 692

4. 67×7=?
   A) 469   B) 459   C) 369   D) 349

5. 86×9=?
   A) 774   B) 764   C) 754   D) 744

6. 97×7=?
   A) 669   B) 679   C) 689   D) 699

7. 74×8=?
   A) 482   B) 492   C) 592   D) 562

8. 79×9=?
   A) 611   B) 612   C) 710   D) 711

9. 88×8=?
   A) 614   B) 664   C) 694   D) 704

10. 55×9=?
    A) 495   B) 485   C) 487   D) 505

TEST – 48

1. 67×17=?
   A) 1139   B) 1249   C) 1259   D) 1279

2. 64×22=?
   A) 1208   B) 1308   C) 1408   D) 1318

3. 98×18=?
   A) 1764   B) 1664   C) 1564   D) 1982

4. 92×92=?
   A) 8454   B) 8354   C) 8464   D) 8463

5. 73×23=?
   A) 1579   B) 1479   C) 1679   D) 1689

6. 44×33=?
   A) 1352   B) 1452   C) 1652   D) 1792

7. 78×78=?
   A) 6084   B) 6074   C) 6094   D) 7074

8. 59×19=?
   A) 1221   B) 1321   C) 1421   D) 1121

9. 64×29=?
   A) 1847   B) 1856   C) 1746   D) 1546

10. 49×39=?
    A) 1911   B) 1811   C) 1711   D) 1611

TEST – 49
(Estimate)

1. 23+52+63=?
   A) 110   B) 120   C) 130   D) 140

2. 71+81+99=?
   A) 210   B) 230   C) 240   D) 250

3. 23+43+93=?
   A) 130   B) 140   C) 150   D) 160

4. 27+37+47=?
   A) 110   B) 120   C) 130   D) 140

5. 24+57+64=?
   A) 130   B) 140   C) 150   D) 160

6. 71+75+77=?
   A) 230   B) 250   C) 240   D) 270

7. 91+97+98=?
   A) 280   B) 290   C) 270   D) 260

8. 13+19+27=?
   A) 130   B) 140   C) 60   D) 160

9. 98+78+28+18=?
   A) 210   B) 220   C) 225   D) 230

10. 23+43+53+99=?
   A) 180   B) 190   C) 200   D) 210

TEST – 50

1. $175 \div 5 = ?$
   A) 25   B) 35   C) 37   D) 45

2. $636 \div 6 = ?$
   A) 102   B) 103   C) 104   D) 106

3. $995 \div 5 = ?$
   A) 188   B) 177   C) 166   D) 199

4. $1964 \div 11 = ?$
   A) 179   B) 144   C) 154   D) 164

5. $1452 \div 22 = ?$
   A) 33   B) 44   C) 55   D) 66

6. $777 \div 3 = ?$
   A) 239   B) 249   C) 259   D) 269

7. $968 \div 11 = ?$
   A) 88   B) 77   C) 78   D) 98

8. $1024 \div 8 = ?$
   A) 122   B) 124   C) 128   D) 129

9. $289 \div 17 = ?$
   A) 12   B) 16   C) 17   D) 27

10. $955 \div 5 = ?$
    A) 161   B) 171   C) 181   D) 191

TEST – 51

1. $19^2=?$
   A) 351   B) 361   C) 381   D) 381

2. $22^2=?$
   A) 386   B) 484   C) 374   D) 394

3. $23^2=?$
   A) 429   B) 529   C) 549   D) 269

4. $32^2=?$
   A) 1024   B) 1044   C) 1064   D) 1084

5. $18^2+8^2=?$
   A) 384   B) 386   C) 387   D) 388

6. $14^2+4^2=?$
   A) 212   B) 213   C) 214   D) 216

7. $15^2+5^2=?$
   A) 230   B) 235   C) 250   D) 255

8. $17^2+7^2=?$
   A) 318   B) 328   C) 338   D) 348

9. $21^2+11^2=?$
   A) 362   B) 462   C) 562   D) 572

10. $11^2+12^2+13^2+14^2=?$
    A) 620   B) 624   C) 628   D) 630

TEST – 52

1. 64×50=?
   A) 3100
   B) 3200
   C) 3400
   D) 3600

2. 98×50=?
   A) 4900
   B) 1800
   C) 4700
   D) 4600

3. 77×40=?
   A) 3060
   B) 3070
   C) 3080
   D) 3090

4. 67×60=?
   A) 4020
   B) 4030
   C) 4050
   D) 4070

5. 88×80=?
   A) 7010
   B) 7020
   C) 7030
   D) 7040

6. 440×40=?
   A) 17700
   B) 17650
   C) 17750
   D) 17600

7. 96×500=?
   A) 47000
   B) 4800
   C) 48000
   D) 4700

8. 628×50=?
   A) 3140
   B) 3240
   C) 33400
   D) 31400

9. 45×50=?
   A) 3140
   B) 2150
   C) 2240
   D) 2250

10. 75×70=?
    A) 5250
    B) 5350
    C) 5450
    D) 5550

TEST – 53
(Estimate)

1. 135+327+644=?
   A) 900   B) 1000
   C) 1200   D) 1300

2. 145+215+775=?
   A) 100   B) 1100
   C) 1200   D) 1300

3. 115+295+335=?
   A) 700   B) 400
   C) 500   D) 600

4. 225+335+555=?
   A) 800   B) 900
   C) 1000   D) 1100

5. 665+777+999=?
   A) 2000   B) 2100
   C) 2500   D) 2400

6. 105+205+395=?
   A) 500   B) 600
   C) 700   D) 800

7. 545+743+842=?
   A) 1700   B) 1800
   C) 1900   D) 2000

8. 905+855+455=?
   A) 1000   B) 1100
   C) 2300   D) 1300

9. 715-415+375=?
   A) 700   B) 600
   C) 500   D) 300

10. 1465+1205+1185=?
    A) 3600   B) 3700
    C) 3800   D) 3000

TEST – 54

1. $8 \times 36 \div 6 = ?$
   A) 38   B) 48   C) 52   D) 42

2. $90 \times 90 \div 10 = ?$
   A) 810   B) 720   C) 710   D) 920

3. $44 \times 44 \div 11 = ?$
   A) 166   B) 176   C) 186   D) 196

4. $77 \times 55 \div 11 = ?$
   A) 355   B) 345   C) 385   D) 375

5. $22 \times 22 \div 2 = ?$
   A) 212   B) 213   C) 228   D) 242

6. $(6 + 6 \div 6) + (36 + 36 \div 6) = ?$
   A) 49   B) 42   C) 43   D) 44

7. $(100 \times 100 \div 10) + (9 \times 9 \div 9) = ?$
   A) 1009   B) 1011
   C) 1012   D) 1013

8. $6 \times 50 \div 5 = ?$
   A) 60   B) 70   C) 80   D) 90

9. $8 \times 80 \div 16 = ?$
   A) 30   B) 40   C) 50   D) 60

10. $3^2 \times 3^2 \div 3 = ?$
    A) 12   B) 13   C) 14   D) 27

TEST – 55

1. If 1 pen=0.25 C then 2 dozen pens=$
   A) 6　　　B) 7　　　C) 8　　　D) 9

2. If 1 pencil=0.35 C then 3 dozen pencils=$
   A) 12.8　B) 12.6　C) 12.7　D) 12.9

3. If 3 algebra books=$12.12, then 9 algebra books=$
   A) 36　　　　　　B) 36.24
   C) 36.36　　　　D) 38.36

4. If 4 geometry books=$16.16 then one dozen geometry books=$
   A) 48　　　　　　B) 48.32
   C) 48.48　　　　D) 48.68

5. If 6 marbles=$7.2, then 18 marbles=$
   A) 21.6　　　　　B) 43.1
   C) 43.2　　　　　D) 43.18

6. If 3 apples=$3.12, then 18 apples=$
   A) 18　　　　　　B) 18.22
   C) 18.62　　　　D) 18.72

7. If one dozen pens=$24.24 find one pen price.
   A) $2　　B) $2.4　C) $2.5　D) $2.2

8. If one dozen eggs=$36.24 find 3 eggs price.
   A) $8.6　B) $7.6 C) $8.12 D) $9.06

9. If 3 watermelon=$18.72 then 2 watermelon=$
   A) $12　　　　　　B) $12.38
   C) $12.40　　　　D) $12.48

10. If two dozen lemon=14.4 then 3 lemons=$
    A) 1　　　B) 1.2　　C) 1.6　　D) 1.8

TEST – 56

1. How many primes numbers between 10 and 30
   A) 6   B) 7   C) 8   D) 9

2. How many primes numbers between 2 and 30?
   A) 7   B) 8   C) 9   D) 10

3. How many composite number between 7 and 27?
   A) 11   B) 12   C) 13   D) 14

4. How many composite number between 3 and 13?
   A) 3   B) 4   C) 5   D) 6

5. How many primes number less than 28?
   A) 6   B) 7   C) 8   D) 9

6. How many primes number less than 42?
   A) 13   B) 10   C) 11   D) 12

7. How many primes number less than 60?
   A) 15   B) 16   C) 17   D) 18

8. Sum all primes number less than 25.
   A) 100   B) 102   C) 104   D) 106

9. Sum all primes numbers less than 50.
   A) 312   B) 316   C) 318   D) 328

10. Sum all composite numbers less than 17.
    A) 94   B) 95   C) 96   D) 97

TEST – 57

1. 984 ÷ 4 = ?
   A) 234   B) 246   C) 256   D) 276

2. 996 ÷ 8 = ?
   A) 124   B) 124.5   C) 123   D) 123.5

3. 664 ÷ 4 = ?
   A) 156   B) 166   C) 176   D) 186

4. 7784 ÷ 4 = ?
   A) 1646   B) 1746   C) 1846   D) 1946

5. 7791 ÷ 3 = ?
   A) 2597   B) 2487   C) 2397   D) 2097

6. 8558 ÷ 11 = ?
   A) 768   B) 778   C) 788   D) 798

7. 5575 ÷ 7 = ?
   A) 625   B) 725   C) 745   D) 796.4

8. 6909 ÷ 7 = ?
   A) 947   B) 967   C) 987   D) 917

9. 30199 ÷ 23 = ?
   A) 1313   B) 1213   C) 1214   D) 1327

10. 3663 ÷ 11 = ?
    A) 111   B) 222   C) 333   D) 343

TEST – 58
(Estimate)

1. 12×21×37=?
   A) 800   B) 8000
   C) 700   D) 7000

2. 21×42×53=?
   A) 40   B) 400
   C) 4000   D) 40000

3. 97×87×61=?
   A) 540000   B) 54000
   C) 5400   D) 540

4. 13×23×73=?
   A) 140   B) 1400
   C) 14000   D) 15000

5. 19×18×17=?
   A) 8000   B) 800
   C) 80   D) 7000

6. 14×25×36=?
   A) 1200   B) 12000
   C) 1400   D) 14000

7. 94×84×74=?
   A) 640000   B) 50600
   C) 504000   D) 506000

8. 13×23×33=?
   A) 500   B) 600
   C) 5000   D) 6000

9. 25×35×45=?
   A) 5000   B) 50000
   C) 6000   D) 60000

10. 12×21×66=?
    A) 1400   B) 14000
    C) 1200   D) 12000

COMPOSITE NUMBERS

Composite N=4, 6, 8, 9, 10, 12, 14, 15

TEST – 59

1. How many composite number less than 17?
   A) 8   B) 9   C) 10   D) 11

2. How many composite numbers less than 33?
   A) 17   B) 18   C) 19   D) 20

3. How many composite numbers between 17 and 29?
   A) 8   B) 9   C) 10   D) 11

4. Sum all composite numbers between 3 and 17.
   A) 93   B) 94   C) 95   D) 96

5. Sum all composite numbers between 19 and 29.
   A) 192   B) 193   C) 194   D) 195

6. Which number is composite number?
   A) 19   B) 23   C) 29   D) 32

7. Which number is composite number?
   A) 31   B) 33   C) 43   D) 73

8. How many composite numbers less than 23?
   A) 10   B) 11   C) 12   D) 13

9. How many composite numbers less than 31?
   A) 19   B) 20   C) 21   D) 22

10. Set
    A:{9,10,17,19,23,31,44,47,53,55}
    How many composite numbers are set A?
    A) 4   B) 5   C) 6   D) 7

TEST – 60

Even numbers: 2,4,6,8,10,12,…
Odd numbers: 1,3,5,7,9,11,…

1. How many even numbers between 5 and 25?
    A) 10   B) 11   C) 12   D) 16

2. How many even numbers between 31 and 61?
    A) 16   B) 15   C) 18   D) 19

3. How many even numbers between 21 and 91?
    A) 32   B) 33   C) 35   D) 36

4. How many even number between 121 and 391?
    A) 130   B) 131   C) 132   D) 133

5. How many odd number between 2 and 32?
    A) 14   B) 15   C) 16   D) 17

6. How many odd numbers between 24 and 92?
    A) 35   B) 40   C) 41   D) 42

7. How many odd numbers between 52 and 152?
    A) 51   B) 72   C) 75   D) 76

8. How many odd numbers between 102 and 214?
    A) 67   B) 66   C) 65   D) 57

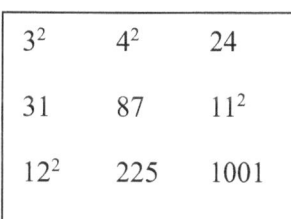

3^2	4^2	24
31	87	11^2
12^2	225	1001

Set A

9. How many odd numbers at set A?
    A) 8   B) 7   C) 6   D) 5

10. How many even numbers at set A?
    A) 6   B) 5   C) 4   D) 3

TEST – 61

(Cube root)

Numbers	1	2	3	4	5	6	7	8	9	10
Cubes	1	8	27	64	125	216	343	512	729	1000

1. What is the cube root of 1331?
   A) 11   B) 12   C) 18   D) 19

2. What is the cube root of 1728?
   A) 13   B) 12   C) 11   D) 10

3. What is the cube root of 1000?
   A) 10   B) 11   C) 20   D) 100

4. What is the cube root of 3375?
   A) 13   B) 14   C) 15   D) 25

5. What is the cube root of 512?
   A) 7   B) 8   C) 9   D) 10

6. What is the cube root of 8000?
   A) 10   B) 20   C) 40   D) 80

7. What is the cube root of 2700?
   A) 90   B) 60   C) 30   D) 13

8. What is the cube root of 216000?
   A) 30   B) 40   C) 50   D) 60

9. What is the cube root of 125000?
   A) 20   B) 30   C) 50   D) 500

10. What is the cube root of $\dfrac{64}{216}$?

   A) $\dfrac{4}{6}$ B) $\dfrac{6}{4}$ C) $\dfrac{8}{6}$ D) $\dfrac{6}{8}$

TEST – 62
(Standard form)

Write each numbers standard form.

1. $23 \times 15 = ?$
   A) 23   B) 123
   C) 230   D) 345

2. $44 \times 10^2 = ?$
   A) 4400   B) 440
   C) 4.4   D) 0.44

3. $6.7 \times 10^3 = ?$
   A) 67000   B) 6700
   C) 670   D) 67

4. $0.45 \times 10^4 = ?$
   A) 4500   B) 450
   C) 45   D) 4.4

5. $9.7 \times 10^5 = ?$
   A) 97   B) 970
   C) 9700   D) 970000

6. $8.4 \times 10^4 = ?$
   A) 840   B) 8400
   C) 84000   D) 840000

7. $29 \times 10^3 = ?$
   A) 290   B) 2900
   C) 2900   D) 29000

8. $14 \times 10^3 + 16 \times 10^3 = ?$
   A) 280   B) 2800
   C) 28000   D) 30000

9. $1.7 \times 10^4 + 7.1 \times 10^4 = ?$
   A) 88   B) 880
   C) 8800   D) 88000

10. $1.3 \times 10^3 + 1.7 \times 10^3 + 30 \times 10^2 = ?$
    A) 300   B) 3000
    C) 6000   D) 600

PRIME FACTORIZATION

The prime factorization of a number is the number written as the product of its prime factors.

Example: The prime factorization of 21 is: 3, 7

Example: The prime factorization of 28 is: $2 \cdot 2 \cdot 7 = 2^2 \cdot 7$

Example: The prime factorization of 300 is: $3 \cdot 2 \cdot 2 \cdot 5 \cdot 5$ or $3 \cdot 2^2 \cdot 5^2$.

TEST – 63

Find the prime factorization of each number

1. 24
   A) $4 \cdot 6$ B) $8 \cdot 3$
   C) $12 \cdot 2$ D) $2 \cdot 2 \cdot 2 \cdot 3$

2. 44
   A) $2 \cdot 2 \cdot 11$ B) $4 \cdot 11$
   C) $2 \cdot 22$ D) $1 \cdot 44$

3. 78
   A) $6 \cdot 13$ B) $2 \cdot 39$
   C) $2 \cdot 3 \cdot 13$ D) $13 \cdot 7$

4. 84
   A) $12 \cdot 7$ B) $14 \cdot 6$
   C) $4 \cdot 21$ D) $7 \cdot 2 \cdot 2 \cdot 3$

5. 120
   A) $8 \cdot 15$ B) $2^3 \cdot 3^1 \cdot 5^1$
   C) $4 \cdot 6 \cdot 5$ D) $18 \cdot 9$

6. 400
   A) $2^4 \cdot 5^2$ B) $4 \cdot 100$
   C) $25 \cdot 16$ D) $25 \cdot 8$

7. 600
   A) $6 \cdot 100$ B) $2^3 \cdot 3^1 \cdot 5^2$
   C) $3^2 \cdot 2^1 \cdot 5^2$ D) $12 \cdot 50$

8. 1500
   A) $15 \cdot 100$ B) $5 \cdot 300$
   C) $4 \cdot 3 \cdot 5^3$ D) $2^2 \cdot 3 \cdot 5^3$

9. 77
   A) $7 \cdot 11$ B) $1 \cdot 77$
   C) $77 \cdot 2$ D) $22 \cdot 6$

10. 97
    A) $32 \cdot 3$ B) $27 \cdot 3$
    C) $97 \cdot 1$ D) $33 \cdot 3$

RECIPROCALS

Two numbers whose product is 1 are reciprocals.

Example: The reciprocal of $\frac{3}{7}$ is $\frac{7}{3}$.   Example: The reciprocal of 9 is $\frac{1}{9}$.

Example: The reciprocal of $2\frac{3}{4}$ is $\frac{4}{11}$.

TEST – 64

Find the reciprocal of each number.

1. $\frac{7}{15}$

   A) $\frac{15}{7}$ B) $-\frac{7}{15}$ C) $-\frac{15}{7}$ D) $\frac{3}{6}$

2. $\frac{2}{11}$

   A) $-\frac{2}{11}$ B) $\frac{11}{2}$ C) $-\frac{11}{2}$ D) $\frac{1}{5}$

3. $\frac{4}{9}$

   A) $-\frac{4}{9}$ B) $-\frac{9}{4}$ C) $\frac{9}{4}$ D) 1

4. $-\frac{3}{4}$

   A) $\frac{3}{4}$ B) $-\frac{3}{4}$ C) $\frac{4}{3}$ D)
   $-\frac{4}{3}$

5. $-\frac{3}{11}$

   A) $\frac{3}{11}$ B) $-\frac{11}{3}$ C) $-\frac{3}{11}$ D) $3\frac{1}{3}$

6. $3\frac{4}{5}$

   A) $-3\frac{4}{5}$ B) $-\frac{5}{19}$ C) $\frac{5}{19}$ D) $-\frac{19}{4}$

7. $7\frac{3}{4}$

   A) $-7\frac{3}{4}$ B) $4\frac{3}{7}$ C) $\frac{4}{31}$ D) $-\frac{31}{4}$

8. $-6\frac{5}{7}$

   A) $\frac{47}{5}$ B) $-\frac{47}{5}$ C) $\frac{37}{5}$ D) $-\frac{7}{47}$

9. $-\frac{6}{17}$

   A) $\frac{17}{6}$ B) $\frac{6}{17}$ C) $-\frac{17}{6}$ D) $-3\frac{1}{6}$

10. $\frac{11}{23}$

    A) $-\frac{11}{23}$ B) $-\frac{23}{11}$ C) $\frac{23}{11}$ D) $2\frac{4}{11}$

ABSOLUTE VALUE

The absolute value of a number is the number`s distance from 0 on a number line.

|a|=a |-a|=a

Example: |-7|+|-2|=?   Solution: |-7|+|-2|=7+2=9

Example: $\left|-\frac{1}{3}\right|+\left|\frac{5}{3}\right| = ?$ Solution: $\left|-\frac{1}{3}\right|+\left|\frac{5}{3}\right| = \frac{1}{3}+\frac{5}{3}=\frac{6}{3}=2$

TEST – 65

1. |-9|+9=?
   A) 0   B) 9   C) 18   D) -18

6. $\left|\pi - 2\right|+\left|\pi - 2\right| = ?$
   A) 0   B) $2\pi - 2$
   C) $4 - 2\pi$ D) $2\pi - 4$

2. |-4|+|2²|=?
   A) 8   B) 4   C) 6   D) 16

7. |-3²|+|-2³|=?
   A) 15   B) 16   C) 17   D) 18

3. $\left|-\frac{1}{3}\right|+\left|\frac{1}{2}\right| = ?$
   A) $-\frac{1}{6}$ B) $-\frac{1}{5}$ C) $\frac{1}{6}$ D) $\frac{5}{6}$

8. $\left|-2\right|\cdot\left|-\frac{1}{4}\right|\cdot\left|4\right| = ?$
   A) -2   B) 4   C) -4   D) 2

4. |9-4|+|8-5|=?
   A) 7   B) 8   C) 9   D) 10

9. |17-7|+|16-6|=?
   A) 16   B) 17   C) 18   D) 20

5. |12-2|+|2-12|=?
   A) 0   B) -20   C) -24   D) 20

10. $\left|\sqrt{9} - \sqrt{4}\right|+\left|\sqrt{36} - \sqrt{25}\right| = ?$
    A) 0   B) 1   C) 2   D) 3

GEOMETRIC SEQUENCE

1,3,9,27,81,243,....   2,4,8,16,32,64,...

A geometric sequence is sequence of the form $a_1, a_1r, a_1r^2, a_1r^3, a_1r^4, ...$

a_1=first term, r=common ratio

Example:

6, 12, 24, 48, 96,...

6   6·2 6·44 6·8 6·12

a_1 a_2 a_3 a_4 a_5

6,12,24,48,96,...

Each term is found by multiplying the previous term by 2.

TEST – 66

1. What is the next term in the geometric sequence 4, 8, 16, 32, 64, ?
   A) 128   B) 138   C) 148   D) 158

2. What is the next term in the geometric sequence 5,15,45, 135,?
   A) 205   B) 305   C) 315   D) 405

3. What is the next term in the geometric sequence 7,21,63, 252,?
   A) 756   B) 776   C) 786   D) 796

4. Find the next term of geometric sequence 6,18,54,162,?
   A) 486   B) 496   C) 516   D) 576

5. Find the next term of the geometric sequence 12,24,48,96,?
   A) 162   B) 172   C) 182   D) 192

6. Find the next term of the x_1 geometric sequence 7,28,112,448,?
   A) 1592   B) 1692
   C) 1792   D) 1892

7. x_1...3,12,48,192,?
   A) 748   B) 758   C) 768   D) 798

8. x_1...5,20,80,320,?
   A) 1280   B) 1290
   C) 1380   D) 1390

9. x_1...2,10,50,250,?
   A) 1150   B) 1200
   C) 1230   D) 1250

10. x_1...8,24,72,216,?
    A) 618   B) 628   C) 638   D) 648

ESTIMATE

Example: 1320-32×22

Solution: 1000-30×20
1000-600=400

Example: 16650-25×25

Solution: 2000-30×30
2000-900=1100

TEST – 67

1. Estimate: 2142-21×21
    A) 1600   B) 1500
    C) 1400   D) 1200

2. Estimate: 927-22×22
    A) 500 B) 400 C) 300 D) 200

3. Estimate: 3248-33×33
    A) 2000   B) 2100
    C) 2200   D) 2300

4. Estimate: 13250-55×55
    A) 9000   B) 9100
    C) 9200   D) 9400

5. Estimate: 17267-63×61
    A) 16400   B) 16300
    C) 16200   D) 16100

6. Estimate: 212648-624×324
    A) 15200   B) 2000
    C) 20000   D) 182000

7. 2776-35×25
    A) 1800   B) 1700
    C) 1600   D) 1500

8. 3445-17×18
    A) 2100   B) 2200
    C) 2300   D) 2600

9. 1948-19×19
    A) 1300   B) 1400
    C) 1500   D) 1600

10. 2764+41×51
    A) 5000 B) 9000
    C) 8000   D) 7000

TEST – 68
(Rounded to the tenth)

1. 6.294
    A) 6.2   B) 6.3   C) 6.4   D) 6.5

2. 7.793
    A) 7.7   B) 7.8   C) 7.9   D) 8

3. 6.124
    A) 6.2   B) 6.3   C) 6.4   D) 6.1

4. 97.623
    A) 97   B) 97.7   C) 97.8   D) 97.6

5. 32.824
    A) 32.3   B) 32.9   C) 32.8   D) 33.8

6. 89.89
    A) 89   B) 89.9   C) 90   D) 89.8

7. 99.99
    A) 100   B) 99.9   C) 99   D) 99.8

8. 19.97
    A) 20   B) 19   C) 19.9   D) 19.8

9. 199.88
    A) 200   B) 199   C) 201   D) 199.9

10. 1999.1999
    A) 2000   B) 1999
    C) 1999.1   D) 1999.2

THE SQUARE OF ROOT

Example: The square of 225 is 15   (15×15=225)

Example: The square root of 900 is 30   (30×30=900)

Example: The square of 196 is 14   (14×14=196)

TEST – 69

1. The square root of 400 is …
    A) 12   B) 18   C) 19   D) 20

2. The square root of 289 is…
    A) 16   B) 17   C) 18   D) 19

3. The square root of 484 is…
    A) 21   B) 22   C) 23   D) 24

4. The square root of 1225 is…
    A) 25   B) 35   C) 45   D) 55

5. The square root of 1600 is…
    A) 30   B) 40   C) 42   D) 80

6. The square root of 2500 is…
    A) 30   B) 25   C) 45   D) 50

7. The square root of 3600 is…
    A) 30   B) 50   C) 45   D) 60

8. The square root of 10000 is…
    A) 10   B) 100   C) 120   D) 1000

9. The square root of 6400 is…
    A) 80   B) 70   C) 60   D) 40

10. The square root of 529 is…
    A) 20   B) 21   C) 22   D) 23

TEST – 70

1. 4^4=?
   A) 16   B) 64   C) 128   D) 256

2. 3^5=?
   A) 243   B) 15   C) 143   D) 85

3. 2^5=?
   A) 10   B) 16   C) 32   D) 64

4. 5^3=?
   A) 15   B) 25   C) 75   D) 125

5. 10^3+5^3=?
   A) 1000   B) 1125
   C) 1025   D) 1035

6. 3^5+5^3=?
   A) 268   B) 168   C) 468   D) 368

7. $2^3+3^3+4^3$=?
   A) 88   B) 99   C) 109   D) 69

8. 4^4-3^4=?
   A) 175   B) 185   C) 195   D) 205

9. 7^3+5^3=?
   A) 268   B) 368   C) 468   D) 568

10. $2^2+3^2+4^2+5^2$=?
    A) 34   B) 54   C) 64   D) 74

EQUATION

1) cx+a=bb, $x = \dfrac{bb - a}{c}$ Example: 2x+4=22, $x = \dfrac{22 - 4}{2} = \dfrac{18}{2} = 9$

2) ax+b=cd, $x = \dfrac{cd - b}{a}$ Example: 3x+4=25 $x = \dfrac{25 - 4}{3} = \dfrac{21}{3} = 7$

TEST – 71

1. 2x+5=25. What is the x+7=?
    A) 10   B) 12   C) 15   D) 17

2. 3x+3=24. What is the x^2+4=?
    A) 52   B) 53   C) 54   D) 55

3. 5x+6=21. What is the x^2-7=?
    A) 18   B) 19   C) 20   D) 2

4. 7x+7=21. What is the 3x+4=?
    A) 8   B) 9   C) 10   D) 11

5. 10x+44=84. What is the x-4=?
    A) 1   B) 2   C) 0.4   D) 0

6. 6x+6=36. What is the x-3=?
    A) 1   B) 2   C) 3   D) 4

7. If 4x+4=24, what is the x^2+x=?
    A) 20   B) 22   C) 25   D) 30

8. If 2x+8=48, what is the 3x-5=?
    A) 55   B) 65   C) 70   D) 75

9. If 3x+12=54, what is the x-14=?
    A) 14   B) 13   C) 12   D) 0

10. If 4x+5=77, what is the x-14=?
    A) 1   B) 2   C) 3   D) 4

REDUCE $\dfrac{a}{b}$ TO LOWEST FORM

Example: $\dfrac{24}{28} = \dfrac{24 \div 4}{28 \div 4} = \dfrac{6}{7}$ Example: $\dfrac{27}{36} = \dfrac{3 \cdot \cancel{9}}{4 \cdot \cancel{9}} = \dfrac{3}{4}$ Example: $\dfrac{35}{55} = \dfrac{7 \cdot \cancel{5}}{11 \cdot \cancel{5}} = \dfrac{7}{11}$

TEST – 72

1. Reduce $\dfrac{12}{15}$ to lowest form.

   A) $\dfrac{5}{4}$ B) $\dfrac{4}{5}$ C) $\dfrac{3}{4}$ D) $\dfrac{3}{5}$

2. Reduce $\dfrac{15}{27}$ to lowest form.

   A) $\dfrac{3}{5}$ B) $\dfrac{5}{6}$ C) $\dfrac{9}{5}$ D) $\dfrac{5}{9}$

3. Reduce $\dfrac{25}{35}$ to lowest form.

   A) $\dfrac{5}{7}$ B) $\dfrac{7}{5}$ C) $\dfrac{5}{3}$ D) $\dfrac{7}{3}$

4. Reduce $\dfrac{49}{21}$ to lowest form.

   A) $\dfrac{7}{4}$ B) $\dfrac{7}{2}$ C) $\dfrac{3}{7}$ D) $\dfrac{7}{3}$

5. Reduce $\dfrac{42}{36}$ to lowest form.

   A) $\dfrac{7}{8}$ B) $\dfrac{6}{7}$ C) $\dfrac{7}{6}$ D) $\dfrac{7}{9}$

6. Reduce $\dfrac{21}{27}$ to lowest form.

   A) $\dfrac{7}{8}$ B) $\dfrac{7}{6}$ C) $\dfrac{7}{9}$ D) $\dfrac{9}{7}$

7. Reduce $\dfrac{45}{36}$ to lowest form.

   A) $\dfrac{5}{4}$ B) $\dfrac{5}{6}$ C) $\dfrac{4}{5}$ D) $\dfrac{6}{5}$

8. Reduce $\dfrac{25}{20}$ to lowest form.

   A) $\dfrac{5}{4}$ B) $\dfrac{4}{5}$ C) $\dfrac{5}{6}$ D) $\dfrac{6}{7}$

9. Reduce $\dfrac{30}{48}$ to lowest form.

   A) $\dfrac{5}{8}$ B) $\dfrac{5}{6}$ C) $\dfrac{8}{5}$ D) $\dfrac{10}{6}$

10. Reduce $\dfrac{144}{225}$ to lowest form.

    A) $\dfrac{16}{20}$ B) $\dfrac{16}{25}$ C) $\dfrac{20}{25}$ D) $\dfrac{25}{20}$

ADD FRACTION

1) $\dfrac{x}{a} + \dfrac{y}{a} = \dfrac{x+y}{a}$ $\qquad\qquad$ $\dfrac{3}{11} + \dfrac{4}{11} = \dfrac{7}{11}$

2) $\dfrac{x}{a} + \dfrac{y}{a} + \dfrac{z}{a} = \dfrac{x+y+z}{a}$ $\qquad$ $\dfrac{2}{17} + \dfrac{3}{17} + \dfrac{8}{18} = \dfrac{13}{17}$

TEST – 73

1. $\dfrac{3}{7} + \dfrac{4}{7} = ?$

   A) 1 $\qquad$ B) 2 $\qquad$ C) $\dfrac{1}{7}$ $\qquad$ D) $\dfrac{12}{7}$

2. $\dfrac{4}{13} + \dfrac{5}{13} = ?$

   A) $\dfrac{20}{13}$ $\qquad$ B) $\dfrac{13}{20}$ $\qquad$ C) $\dfrac{8}{13}$ $\qquad$ D) $\dfrac{9}{13}$

3. $\dfrac{6}{12} + \dfrac{5}{12} = ?$

   A) $\dfrac{30}{12}$ $\qquad$ B) $\dfrac{1}{12}$ $\qquad$ C) $\dfrac{11}{12}$ $\qquad$ D) 1

4. $\dfrac{5}{11} + \dfrac{7}{11} = ?$

   A) $\dfrac{35}{11}$ $\qquad$ B) $\dfrac{2}{12}$ $\qquad$ C) $\dfrac{12}{11}$ $\qquad$ D) 1

5. $\dfrac{5}{14} + \dfrac{9}{14} = ?$

   A) 1 $\qquad$ B) 2 $\qquad$ C) $\dfrac{4}{14}$ $\qquad$ D) 1

6. $\dfrac{5}{15} + \dfrac{3}{15} + \dfrac{2}{15} = ?$

   A) $\dfrac{9}{15}$ $\qquad$ B) $\dfrac{15}{9}$ $\qquad$ C) $\dfrac{3}{15}$ $\qquad$ D) $\dfrac{2}{3}$

7. $\dfrac{8}{14} + \dfrac{1}{14} + \dfrac{5}{14} = ?$

   A) $\dfrac{13}{14}$ $\qquad$ B) $\dfrac{14}{14}$ $\qquad$ C) $\dfrac{15}{14}$ $\qquad$ D) $\dfrac{16}{14}$

8. $\dfrac{5}{17} + \dfrac{2}{17} + \dfrac{7}{17} = ?$

   A) $\dfrac{9}{14}$ $\qquad$ B) $\dfrac{7}{14}$ $\qquad$ C) $\dfrac{10}{14}$ $\qquad$ D) $\dfrac{14}{17}$

9. $\dfrac{13}{24} + \dfrac{11}{24} - \left(\dfrac{2}{3} + \dfrac{1}{3}\right) = ?$

   A) 1 $\qquad$ B) 0 $\qquad$ C) 2 $\qquad$ D) $\dfrac{3}{12}$

10. $\dfrac{18}{18} + \dfrac{8}{18} - \left(\dfrac{2}{7} + \dfrac{5}{7}\right) = ?$

   A) 0 $\qquad$ B) 4/9 $\qquad$ C) 2 $\qquad$ D) $\dfrac{2}{3}$

SQUARE

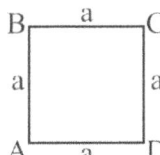

Perimeter=4×a   Area=a×a=a²

Example:

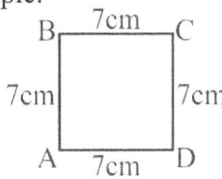

Perimeter=4×7=28 cm   Area=7×7=49 cm²

TEST – 74

1. What is the perimeter of a square with side length 19 sm?
    A) 321   B) 221   C) 76   D) 361

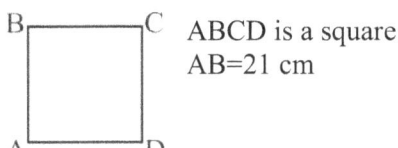

ABCD is a square.
AB=21 cm

2. Find the perimeter.
    A) 421   B) 221   C) 84   D) 88

3. Find the area of square.
    A) 421   B) 431   C) 84   D) 441

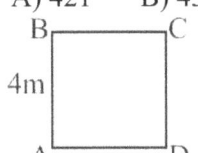

ABCD is a square.
AB=4m cm

4. Find the perimeter.
    A) 8m   B) 12m   C) 16m   D) 16m²

5. Find the area of square.
    A) 8m²   B) 12m²   C) 16m²   D) 20m²

ABCD is a square.

6. Find the perimeter of square.
    A) 4n+8   B) n+8
    C) n+10   D) 4n+2

7. What is the perimeter of square area of 144 cm²?
    A) 36   B) 48   C) 50   D) 64

8. What is the perimeter of square area of 121 cm²?
    A) 22   B) 33   C) 36   D) 44

9. What is the area of square perimeter 80?
    A) 200   B) 300   C) 400   D) 600

10. What is the area of square perimeter 72 cm?
    A) 324 cm²   B) 314 cm²
    C) 296 cm²   D) 196 cm²

THE VOLUME OF CUBE

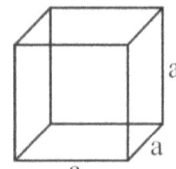

$V = a \times a \times a$
or
$V = a^3$

Example:

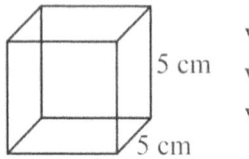

5 cm
5 cm
5 cm

$V = a \times a \times a$
$V = 5 \times 5 \times 5$
$V = 125 \text{ cm}^3$

Example:

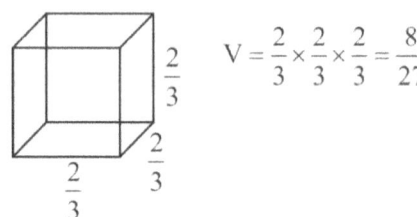

$\frac{2}{3}$
$\frac{2}{3}$
$\frac{2}{3}$

$V = \frac{2}{3} \times \frac{2}{3} \times \frac{2}{3} = \frac{8}{27}$

TEST – 75

1. Find the volume of cube with side length 6 cm.
   A) 206   B) 216   C) 226   D) 236

2. Find the volume of cube with side length 7 cm.
   A) 243 cm³   B) 253 cm³
   C) 343 cm³   D) 353 cm³

3. Find the volume of cube with side length 10 cm.
   A) 100 cm³   B) 100 cm²
   C) 1000 cm³   D) 1000 cm²

4. Find the volume of cube with side length 8 cm.
   A) 512 cm³   B) 412 cm³
   C) 402 cm³   D) 312 cm³

5. Find the volume of cube with side length 11 cm.
   A) 1121   B) 1221
   C) 1321   D) 1331

6. Find the volume of cube with side length $\frac{2}{5}$ cm.
   A) $\frac{4}{25}$ B) $\frac{4}{125}$ C) $\frac{8}{125}$ D) $\frac{8}{145}$

7. Find the volume of cube with side length $\frac{3}{4}$ cm.
   A) $\frac{27}{8}$ B) $\frac{27}{6}$ C) $\frac{27}{64}$ D) $\frac{64}{27}$

8. Find the volume of cube with side length $\frac{1}{10}$ cm.
   A) $\frac{1}{1000}$ B) $\frac{1}{1000}$ C) $\frac{1}{100}$ D) 100

9. Find the volume of cube with side length 0.2 cm.
   A) 0.8   B) 0.08   C) 0.008   D) 0.16

10. Find the volume of cube with side length 0.4 cm.
    A) 6.4   B) 0.64   C) 0.064   D) 64

TRAPEZOID

Example:

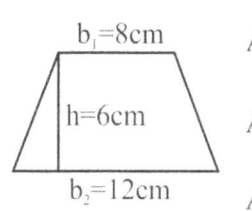

$$Area = \frac{(b_1 + b_2) \cdot h}{2}$$

$$A = \frac{(b_1 + b_2) \cdot h}{2}$$

$$A = \frac{(12 + 8) \cdot 2}{2}$$

$$A = \frac{20 \cdot 6}{2} = 60cm^2$$

b_1 and b_1 base, h height.

TEST – 76

1. A trapezoid with bases of 12 and 16 cm and height of 8 cm. Find the area.
   A) 112   B) 113   C) 114   D) 116

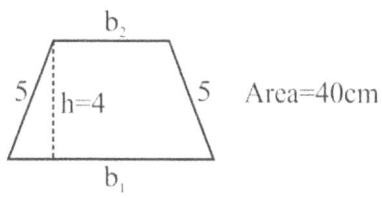

Area=40cm

2.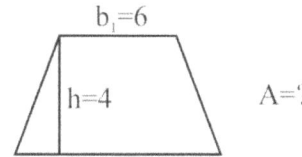
   A) 26   B) 28   C) 30   D) 40

6. $b_1 + b_2 = ?$
   A) 10   B) 20   C) 21   D) 22

7. Perimeter=?
   A) 30   B) 32   C) 33   D) 34

3. A trapezoid with bases 6m and 4m and height 2m. Find the area.
   A) 10m   B) 9m   C) 8m   D) 6m

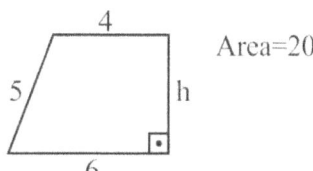

Area=20

4. 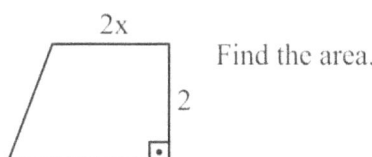 Find the area.
   A) 6x   B) 7x   C) 8x   D) 9x

8. h=?
   A) 3   B) 4   C) 5   D) 6

9. Perimeter=?
   A) 19   B) 20   C) 21   D) 22

5. A trapezoid with bases of 0.8 and 0.6 and height 0.2. Find the area.
   A) 14   B) 1.4   C) 0.14   D) 2.8

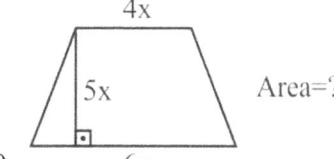

Area=?

10. A) 25   B) 25x   C) $25x^2$ D) $25x^3$

RECTANGULAR

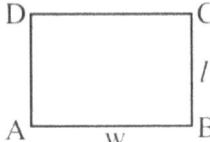

Perimeter=2l+2w
Area=l×w

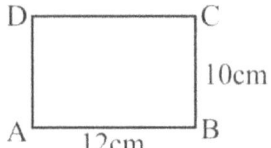

10cm

12cm

Perimeter=2×12+2×10=44cm
Example: Area=12×10=120cm^2

TEST – 77

1. The area of a rectangular with length 12 cm and 9 cm is…
   A) 108 cm
   B) 108 cm^2
   C) 98 cm^2
   D) 98 cm

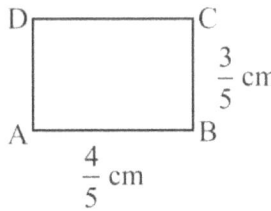

$\frac{3}{5}$ cm

$\frac{4}{5}$ cm

2. Find the perimeter.
   A) 2.2   B) 2.4   C) 2.6   D) 2.8

3. Find the area.
   A) $\frac{12}{35}$ B) $\frac{7}{25}$ C) $\frac{12}{15}$ D) $\frac{12}{25}$

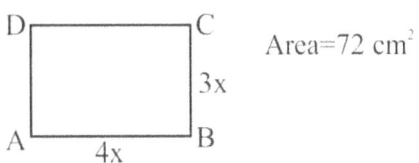

Area=72 cm^2

3x

4x

4. x=?
   A) 1   B) 2   C) 3   D) $\sqrt{6}$

5. Perimeter=?
   A) 30   B) 32   C) 34   D) 14$\sqrt{6}$

ABCD is rectangle. AB:BC=4:3

Perimeter is 56 cm.

6. AB=?
   A) 12   B) 16   C) 18   D) 20

7. Area=?
   A) 144   B) 160   C) 192   D) 288

8. BC=?
   A) 10   B) 11   C) 12   D) 14

9. Area=?
   A) 392   B) 243   C) 244   D) 266

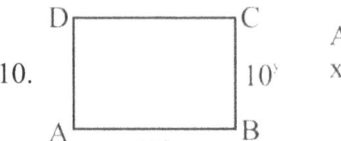

Area=10000
x+y=?

10^y

10^x

10.
   A) 1   B) 2   C) 3   D) 4

TEST – 78
(Ratio and rates)

1. Which rates is unit rate?

   A) $\dfrac{\$100}{5 \text{ ticket}}$ B) $\dfrac{20 \text{ km}}{20 \text{ hour}}$

   C) $\dfrac{\$160}{4 \text{ book}}$ D) $\dfrac{40 \text{ gallon}}{\text{every hours}}$

2. Which is the difference?

   A) 4 to 5   B) $\dfrac{4}{5}$ C) 4:5   D) 5:4

3. Which ratio does not belong?

   A) $\dfrac{4}{6}$ B) $\dfrac{12}{18}$ C) $\dfrac{20}{24}$ D) 6/9

4. 12 miles in 40 min
   36 miles in x min
   A) 80   B) 90   C) 100   D) 120

5. Write the unit rate in fraction forms.
   $600 in 40 days.
   A)15/1day B)16 /1day C)17/day D)20/day

6. Ronaldo walked 8 miles in 3 hours. At the rate how many miles can Ronaldo walk in 9 hours?
   A) 12   B) 18   C) 20   D) 24

7. Jack earned $60 in 4 hours. How much will Jack earn in 16 hours?
   A) $200   B) $240
   C) $260   D) $280

8. Which is the different?

   A) $\dfrac{5}{7}$ B) $\dfrac{15}{21}$ C) $\dfrac{20}{35}$ D) $\dfrac{25}{35}$

9. 16 miles in 50 min
   x   150 min
   A) 48 miles   B) 50 miles
   C) 60 miles   D) 72 miles

10. Ronaldo earned $75 in 5 hours. How much will Ronaldo earn in 15 hours?
    A) $215   B) $225
    C) $245   D) $265

REPEATING DECIMAL FRACTION

- $0.\overline{a} = \dfrac{a}{9}$ $0.\overline{2} = \dfrac{2}{9}$

- $0.\overline{ab} = \dfrac{ab}{99}$ $0.\overline{23} = \dfrac{23}{99}$

- $0.\overline{ab} = \dfrac{ab - a}{90}$ $0.\overline{21} = \dfrac{21 - 2}{90}$

TEST – 79

1. Which is repeating decimal?

   A) $\dfrac{1}{3}$ B) $\dfrac{2}{5}$ C) $\dfrac{3}{5}$ D) $\dfrac{3}{4}$

2. Which is repeating decimal?

   A) $0.\overline{22}$ B) 0.2   C) 0.3   D) 0.55

3. $3.\overline{4} = ?$

   A) $3\dfrac{4}{5}$ B) $3\dfrac{4}{9}$ C) $4\dfrac{3}{9}$ D) $\dfrac{34}{10}$

4. $0.66666… = ?$

   A) $\dfrac{6}{10}$ B) $\dfrac{10}{6}$ C) $\dfrac{9}{6}$ D) $\dfrac{2}{3}$

5. $0.\overline{3} + 0.\overline{4} = ?$

   A) $\dfrac{7}{10}$ B) $\dfrac{7}{9}$ C) $\dfrac{9}{7}$ D) $\dfrac{9}{10}$

6. $0.\overline{75} = ?$

   A) $\dfrac{66}{90}$ B) $\dfrac{65}{90}$ C) $\dfrac{68}{90}$ D) $\dfrac{75}{99}$

7. $0.\overline{243} = ?$

   A) $\dfrac{243}{10}$ B) $\dfrac{243}{100}$ C) $\dfrac{243}{1000}$ D) $\dfrac{243}{999}$

8. $0.24959595… = ?$

   A) $0.24\overline{95}$ B) $0.2\overline{495}$

   C) $0.\overline{249}$ D) $0.249\overline{5}$

9. $0.7555… = ?$

   A) $0.7\overline{5}$ B) 0.75 C) $0.\overline{75}$ D) $0.\overline{75}$

10. $3.777… = ?$

    A) 3.7 B) 37 C) $3.\overline{7}$ D) $0.\overline{37}$

ARITHMETIC MEAN (AVERAGES)

Example: Find the arithmetic mean of 8 and 12.

Solution: $A = \dfrac{x_1 + x_2}{2} = \dfrac{8 + 12}{2} = \dfrac{20}{12} = 10$

Example: Find the arithmetic mean of 10, 12, -8.

Solution: $A = \dfrac{x_1 + x_2 + x_3}{3} = \dfrac{12 + 10 - 8}{3} = \dfrac{14}{3}$

TEST −80

1. Find the average of 18 and 24.
   A) 20   B) 21   C) 22   D) 23

2. Find the average of 4, 8 and 24.
   A) 10   B) 11   C) 12   D) 14

3. Find the average of 6, 14 and -5.
   A) 4   B) 5   C) 6   D) 7

4. Find the average of 6, 8, 12 and 18.
   A) 9   B) 10   C) 11   D) 12

5. 12, 14 and m average is 20. Find m.
   A) 21   B) 22   C) 23   D) 34

6. 5 and m average is 8. Find m.
   A) 11   B) 12   C) 13   D) 14

7. Find the average of $\dfrac{4}{11}$, $\dfrac{5}{11}$ and $\dfrac{2}{11}$.
   A) 1   B) $\dfrac{1}{2}$ C) $\dfrac{1}{3}$ D) $\dfrac{11}{3}$

8. Find the average of $\dfrac{1}{2}$, $\dfrac{1}{3}$ and $\dfrac{1}{6}$.
   A) 1   B) $\dfrac{1}{3}$ C) $\dfrac{1}{4}$ D) $\dfrac{1}{6}$

9. Find the average of 8, 9, 12 and 20.
   A) $\dfrac{49}{4}$ B) $\dfrac{49}{3}$ C) $\dfrac{49}{5}$ D) $\dfrac{47}{4}$

10. Find the average of 6, 8, 12, 18 and -8.
    A) $\dfrac{36}{5}$ B) $\dfrac{37}{5}$ C) $\dfrac{35}{6}$ D) 6

FACTORIALS

$n! = 1 \cdot 2 \cdot 3 \cdot \ldots \cdot n$ $0! = 1$ $1! = 1$ $2! = 1 \cdot 2 = 2$ $3! = 1 \cdot 2 \cdot 3 = 6$ $4! = 1 \cdot 2 \cdot 3 \cdot 4 = 24$

Example: $2! + 3! = 1 \cdot 2 + 1 \cdot 2 \cdot 3 = 2 + 6 = 8$ Example: $8! \div 6! = \dfrac{8 \cdot 7 \cdot \cancel{6!}}{\cancel{6!}} = 8 \cdot 7 = 56$

TEST −81

1. $3! + 8! = ?$
   A) 1208   B) 12212   C) 1239   D) 40326

2. $1! + 2! + 3! + 4! = ?$
   A) 30   B) 31   C) 32   D) 33

3. $(5! + 6!) \div (4! + 3!) = ?$
   A) 28   B) 29   C) 30   D) 34

4. $9! \div 7! = ?$
   A) 70   B) 72   C) 64   D) 60

5. $8! \div (6! + 8!) = ?$
   A) $\dfrac{56}{57}$ B) $\dfrac{57}{56}$ C) $\dfrac{55}{57}$ D) $\dfrac{55}{56}$

6. $\dfrac{5! + 4!}{5!} = ?$
   A) $\dfrac{6}{7}$ B) $\dfrac{7}{5}$ C) $\dfrac{6}{5}$ D) $\dfrac{7}{6}$

7. $8! - 6! = ?$
   A) $8! \cdot 55$ B) $6! \cdot 57$ C) 55 D) $55 \cdot 6!$

8. $\dfrac{6! + 5!}{4! + 3!} = ?$
   A) 27   B) 28   C) 29   D) 30

9. $\dfrac{3! + 2!}{4!} = ?$
   A) $\dfrac{1}{2}$ B) $\dfrac{1}{3}$ C) $\dfrac{1}{4}$ D) $\dfrac{1}{8}$

10. $\dfrac{7! + 6!}{6!} = ?$
    A) 4   B) 5   C) 6   D) 8

CHANGING TO BASE TEN

Example: Change 75_8 to base ten.

Solution:

$$75$$
$$\hookrightarrow 8^0 \cdot 5 = 5$$
$$\hookrightarrow 8^1 \cdot 7 = 56$$
$$56 + 6 = 61$$

Example: Change 546_5 to base ten.

Solution:

$$546$$
$$\hookrightarrow 5^0 = 1 \cdot 6 = 6$$
$$\hookrightarrow 5^1 = 5 \cdot 4 = 20$$
$$\hookrightarrow 5^2 = 25 \cdot 5 = 125$$
$$6 + 20 + 125 = 151$$

TEST – 82

1. Change 36_5 to base ten.
   A) 20   B) 21   C) 22   D) 26

2. Change 76_4 to base ten.
   A) 30   B) 32   C) 34   D) 36

3. Change 246_2 to base ten.
   A) 20   B) 21   C) 22   D) 24

4. Change 984_8 to base ten.
   A) 344   B) 444   C) 544   D) 145

5. Change 778_6 to base ten.
   A) 300   B) 301   C) 92   D) 306

6. Change 2002_3 to base ten.
   A) 18   B) 19   C) 20   D) 56

7. Change 3003_4 to base ten.
   A) 195   B) 196   C) 197   D) 198

8. Change 604_5 to base ten.
   A) 150   B) 152   C) 154   D) 156

9. Change 79_4 to base ten.
   A) 37   B) 36   C) 35   D) 34

10. Change 67_5 to base ten.
    A) 30   B) 33   C) 37   D) 38

PERIMETER OF TRIANGLE

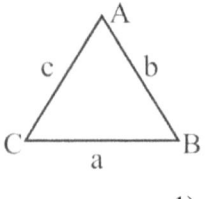

Perimeter=a+b+c

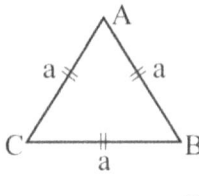

AB=BC=AC
Perimeter=3×a

1)   2)

(equilateral triangle)

TEST −83

1. Perimeter=?

A) 22   B) 24   C) 28   D) 36

2. AB=BC=AC=14 cm
Perimeter=?

A) 42   B) 43   C) 44   D) 48

3. Find the perimeter of an equilateral triangle with side length is 17 cm.
A) 31   B) 41   C) 51   D) 61

4. The equilateral triangle perimeter is 21 cm. Find the one side length.
A) 7   B) 8   C) 9   D) 63

5. The equilateral triangle perimeter is 66 cm. find the one side length.
A) 20   B) 21   C) 22   D) 23

6. x=?

A) 3   B) 16   C) 17   D) 12

7. x=?

A) 27   B) 28   C) 22   D) 18

8.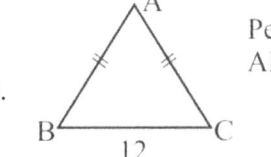
Perimeter=48
AB=AC=?

A) 16   B) 17   C) 18   D) 36

9.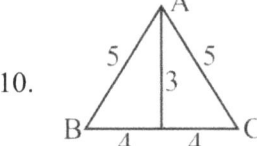
Perimeter=60 cm
BC=24 cm
AB=AC=?

A) 18   B) 19   C) 20   D) 22

10. Sum all triangle perimeter.
A) 39   B) 40   C) 41   D) 18

ANGLE

1)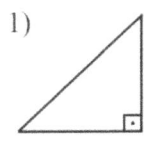
right angle

2)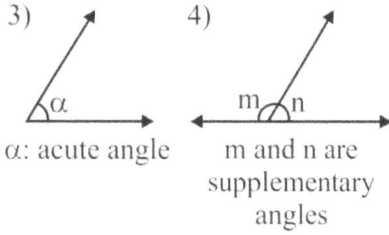
α: obtuse angle

3)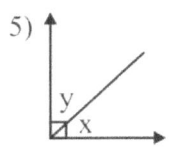
α: acute angle

4)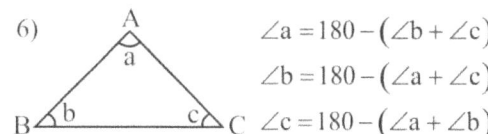
m and n are supplementary angles

5)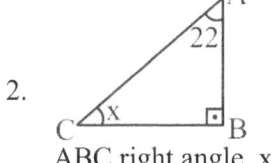
x and y are complementary angles

6)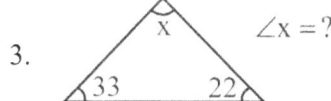
$\angle a = 180 - (\angle b + \angle c)$
$\angle b = 180 - (\angle a + \angle c)$
$\angle c = 180 - (\angle a + \angle b)$

TEST −84

1.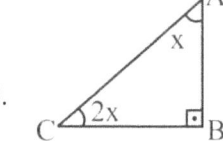
∠x = ?

A) 100°   B) 110°   C) 120°   D) 140°

2.
ABC right angle. x=?

A) 66   B) 68   C) 69   D) 70

3.
∠x = ?

A) 105   B) 115   C) 120   D) 125

4.
x=?

A) 124   B) 114   C) 112   D) 100

5.
x=?

A) 10   B) 20   C) 40   D) 60

6. One complementary angle is 75°. Find the other angle.
A) 15   B) 16   C) 20   D) 25

7. One complementary angle is 122°. Find the other angle.
A) 38   B) 48   C) 58   D) 62

8.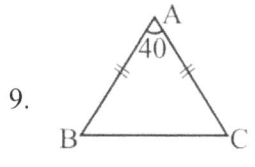
ABC right angle.
∠A = x
∠C = 2x
x = ?

A) 25   B) 30   C) 35   D) 40

9.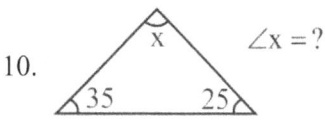
AB = AC
∠A = 40°
∠B = ?

A) 60   B) 70   C) 80   D) 82

10.
∠x = ?

A) 100   B) 110   C) 120   D) 130

TABLE OF COMMON MEASUREMENT

1 minute(mm)=60 seconds   1 hour(hr)=60 minutes   1 day(d)=24 hours(hr)

1 week(wk)=7 days(d)   1 year=365 days(d)   1 year=52 weeks(wk)

1 year=12 months(mo)

TEST −85

1. 2 weeks=… minutes.
   A) 1080   B) 10080
   C) 20160   D) 20080

6. 6 years=… months.
   A) 69   B) 70   C) 71   D) 72

2. 6 months=… weeks.
   A) 12   B) 16   C) 18   D) 24

7. 25 hours=… minutes.
   A) 1300   B) 1400   C) 1500 D) 1160

3. 3 years=… days.
   A) 1089   B) 1091
   C) 1092   D) 1095

8. 8.5 months=… weeks.
   A) 32   B) 34   C) 36   D) 38

4. 4 years=… weeks.
   A) 202   B) 204   C) 206   D) 208

9. 144 hours=… days.
   A) 6   B) 2   C) 8   D) 9

5. 3 days=… hours.
   A) 72   B) 76   C) 86   D) 96

10. 3600 minutes=… days.
    A) 4   B) 3.5   C) 3   D) 2.5

TEST −86

1. 5 years=… weeks.
   A) 240   B) 250   C) 260   D) 280

2. 676 weeks=… years.
   A) 13   B) 14   C) 15   D) 16

3. 4 weeks=… hours.
   A) 90   B) 672   C) 96   D) 106

4. 728 weeks=… years.
   A) 11   B) 12   C) 13   D) 14

5. 504 hours=… weeks.
   A) 2   B) 3   C) 4   D) 5

6. 3.5 hours=… minutes.
   A) 205   B) 210   C) 218   D) 220

7. 7.5 years=… weeks.
   A) 380   B) 385   C) 390   D) 410

8. 1825 days=… years.
   A) 6   B) 5   C) 4   D) 3

9. 147 days=… weeks.
   A) 18   B) 19   C) 20   D) 21

10. 504 hours=… weeks.
    A) 7   B) 6   C) 5   D) 3

TEST −87

1. 13 raised to the second power is…
   A) 26   B) 39   C) 69   D) 169

2. Find the 5 raised to the third power.
   A) 25   B) 50   C) 100   D) 125

3. Find the 6 raised to the third power.
   A) 12   B) 36   C) 72   D) 216

4. Find the 3^2 raised to third power.
   A) 729   B) 81   C) 243   D) 253

5. Find the $\frac{2}{3}$ rasied to the third power.

   A) $\frac{9}{4}$ B) $\frac{27}{8}$ C) $\frac{8}{27}$ D) $\frac{8}{81}$

6. Find the $\frac{4}{5}$ raised to the second power.

   A) $\frac{16}{15}$ B) $\frac{8}{25}$ C) $\frac{16}{25}$ D) $\frac{18}{25}$

7. Find the 4^{-3} raised to the third power.
   A) 4^0 B) 4^1 C) 4^{-6} D) 4^{-9}

8. Find the 15 raised to the forth power.
   A) 40625   B) 43625
   C) 48625   D) 50625

9. Find the 2^2 raised to the fifth power.
   A) 2^4 B) 2^6 C) 2^8 D) 2^{10}

10. 30 raised to the second power is.
    A) 900   B) 600   C) 90   D) 1200

TEST –88

1. $3 + \dfrac{1}{3} + \dfrac{1}{9} = ?$

   A) $\dfrac{31}{3}$ B) $\dfrac{31}{6}$ C) $\dfrac{31}{9}$ D) $\dfrac{32}{7}$

2. $5^1 + 5^2 + 5^3 = ?$
   A) 155   B) 160   C) 170   D) 180

3. $\left(\dfrac{1}{3}\right)^3 + \left(\dfrac{1}{9}\right)^2 = ?$

   A) $\dfrac{5}{81}$ B) $\dfrac{4}{81}$ C) $\dfrac{6}{81}$ D) $\dfrac{7}{81}$

4. $7 + \dfrac{1}{7} + \left(\dfrac{1}{7}\right)^2 = ?$

   A) $\dfrac{161}{49}$ B) $\dfrac{261}{49}$ C) $\dfrac{351}{49}$ D) $\dfrac{361}{49}$

5. $\dfrac{1}{6} + \dfrac{1}{12} + \left(\dfrac{1}{6}\right)^2 = ?$

   A) $\dfrac{5}{18}$ B) $\dfrac{5}{17}$ C) $\dfrac{6}{9}$ D) $\dfrac{7}{36}$

6. $\dfrac{1}{2} + \left(\dfrac{1}{2}\right)^2 + \left(\dfrac{1}{2}\right)^3 = ?$

   A) $\dfrac{7}{8}$ B) $\dfrac{8}{7}$ C) $\dfrac{9}{8}$ D) $\dfrac{11}{8}$

7. $4 + \dfrac{1}{4} + \left(\dfrac{1}{4}\right)^3 = ?$

   A) $\dfrac{215}{64}$ B) $\dfrac{225}{64}$ C) $\dfrac{265}{64}$ D) $\dfrac{273}{64}$

8. $10 + \dfrac{1}{10} + \left(\dfrac{1}{10}\right)^3 = ?$

   A) $\dfrac{10101}{1000}$ B) $\dfrac{10100}{1000}$

   C) $\dfrac{1001}{100}$ D) $\dfrac{10002}{1000}$

9. $8 + \dfrac{1}{8} + \left(\dfrac{1}{2}\right)^3 = ?$

   A) $\dfrac{33}{5}$ B) $\dfrac{33}{6}$ C) $\dfrac{33}{8}$ D) $\dfrac{33}{4}$

10. $2 + \dfrac{1}{2} + \left(\dfrac{1}{2}\right)^2 + \left(\dfrac{1}{2}\right)^3 + \left(\dfrac{1}{2}\right)^4 = ?$

    A) $\dfrac{37}{16}$ B) $\dfrac{93}{16}$ C) $\dfrac{47}{15}$ D) $\dfrac{47}{16}$

TEST –89

1. 1347×11=?
   A) 1307   B) 14817
   C) 1507   D) 1607

2. 367×111=?
   A) 40137   B) 40737
   C) 42737   D) 39137

3. 88×111
   A) 9738   B) 9748
   C) 9758   D) 9768

4. 123×111=?
   A) 13453   B) 13253
   C) 13653   D) 12453

5. 44×3×11=?
   A) 1052   B) 1152
   C) 1252   D) 1452

6. 33×55×11=?
   A) 19765   B) 19865
   C) 19965   D) 18965

7. 14×13×15=?
   A) 2370   B) 2236
   C) 2740   D) 2680

8. 22×33×25=?
   A) 18150   B) 12650
   C) 12720   D) 14620

9. 222×111=?
   A) 24642   B) 24763
   C) 23784   D) 24965

10. Estimate: $(2460×3217)$
    A) 65000   B) 6000
    C) 7500000   D) 75000

TEST –90

1. $36 + \left(36 \div 6\right)^2 = ?$
   A) 62   B) 72   C) 82   D) 92

2. $24 - \left(24 \div 4 + 4\right) = ?$
   A) 14   B) 15   C) 16   D) 18

3. $9 \times 9 \div 3 = ?$
   A) 21   B) 27   C) 81   D) 243

4. $7 + 28 \div 7 + 7 = ?$
   A) 18   B) 24   C) 22   D) 21

5. $18 + 18 \div 9 - 9 = ?$
   A) 21   B) 22   C) 24   D) 11

6. $\left(2 + 8 \div 4\right)^2 + \left(9 + 9 \div 3\right)^2 = ?$
   A) 130   B) 140   C) 150   D) 160

7. 364×10^{-2} write in standard form.
   A) 36400   B) 3640
   C) 3.64   D) 0.364

8. 432×10^{-3} write in standard form.
   A) 432   B) 43.2
   C) 4.32   D) 0.432

9. 0.24×10^{-2} write in standard form.
   A) 2.4   B) 24
   C) 0.024   D) 0.0024

10. 3.6×10^{-3} write in standard form.
    A) 360   B) 36
    C) 0.0036   D) 0.00036

TEST –91

1. Find the remainder of $86 \div 3$.
   A) 6   B) 2   C) 8   D) 9

2. Find the remainder of $98 \div 12$.
   A) 1   B) 2   C) 3   D) 4

3. Find the remainder of $148 \div 12$.
   A) 2   B) 3   C) 4   D) 6

4. Find the remainder of $200 \div 26$.
   A) 11   B) 18   C) 13   D) 14

5. Find the remainder of $97 \div 17$.
   A) 9   B) 8   C) 12   D) 6

6. The product of 2^3 and 3^2 is…
   A) 36   B) 48   C) 64   D) 72

7. Find the product of $\left(\dfrac{1}{2}\right)^2$ and $\left(\dfrac{3}{2}\right)^2$.
   A) 8   B) 9   C) 10   D) 9/16

8. Find the product of $\dfrac{3}{4}$ and $\dfrac{19}{9}$.
   A) 1   B) 2   C) 3   D) $\dfrac{19}{12}$

9. $17^2 + 7^2 = ?$
   A) 218   B) 228   C) 338   D) 248

10. $874 \times 10^3 = ?$
    A) 8740   B) 87400
    C) 874000   D) 847000

TEST –92

1. Estimate: 12+22+32+42=?
    A) 80   B) 90   C) 100   D) 110

2. Estimate: 96+86+76=?
    A) 260   B) 280   C) 290   D) 270

3. Estimate: 91+77+37=?
    A) 210   B) 220   C) 230   D) 240

4. Estimate: 124+265+370=?
    A)   760 B) 900 C) 1000 D) 1200

5. Estimate: 127+166+188=?
    A) 500   B) 600   C) 700   D) 800

6. $32 + 32 \div 32 + 32 = ?$
    A) 66   B) 65   C) 64   D) 63

7. $9 - 9 \div 3 + 3 = ?$
    A) 9   B) 12   C) 14   D) 15

8. $21 - 21 \div 7 + 3 = ?$
    A) 21   B) 22   C) 23   D) 24

9. 1+2+3+...+29=?
    A) 415   B) 425   C) 435   D) 445

10. 2+4+6+...+120=?
    A) 3360   B) 3460
    C) 3560   D) 3660

TEST −93

1. Find the GCF of 12 and 18.
   A) 2   B) 4   C) 6   D) 9

2. Find the GCF of 9 and 24.
   A) 2   B) 3   C) 4   D) 6

3. Find the GCF of 5 and 20.
   A) 5   B) 10   C) 30   D) 40

4. Find the GCF of 5 and 7.
   A) 1   B) 3   C) 10   D) 35

5. Find the GCF of 25 and 75.
   A) 5   B) 10   C) 15   D) 25

6. Find the LCM of (LCM=Least Common Multiple 9 and 21.
   A) 3   B) 7   C) 9   D) 63

7. Find the LCM of 9 and 12.
   A) 18   B) 24   C) 36   D) 42

8. Find the LCM of 5 and 12.
   A) 30   B) 60   C) 65   D) 70

9. Find the LCM of 6 and 15.
   A) 15   B) 20   C) 30   D) 40

10. Find the LCM of 7 and 12.
    A) 84   B) 77   C) 72   D) 48

TEST −94

1. Reduce $\frac{32}{42}$ to lowest form.

   A) $\frac{16}{21}$ B) $\frac{21}{16}$ C) $\frac{8}{12}$ D) $\frac{9}{7}$

2. Find simplify form of $\frac{35}{75}$.

   A) $\frac{7}{12}$ B) $\frac{8}{15}$ C) $\frac{9}{15}$ D) $\frac{7}{15}$

3. Find simplify form of $\frac{27}{48}$.

   A) $\frac{3}{12}$ B) $\frac{3}{8}$ C) $\frac{9}{16}$ D) $\frac{9}{17}$

4. Find simplify form of $\frac{625}{225}$.

   A) $\frac{25}{8}$ B) $\frac{25}{9}$ C) $\frac{15}{9}$ D) $\frac{25}{7}$

5. Find simplify form of $\frac{48}{20}$.

   A) $\frac{12}{5}$ B) $\frac{12}{7}$ C) $\frac{13}{5}$ D) $\frac{13}{7}$

6. 0.9×0.12=?

   A) 1.8   B) 10.8   C) 0.108   D) 1.09

7. 0.22×0.33=?
   A) 7.26   B) 0.726   C) 0.0726   D) 72.6

8. 13^2-12^2=?
   A) 25   B) 26   C) 27   D) 28

9. 18^2-17^2=?
   A) 31   B) 32   C) 34   D) 35

10. 22^2-20^2=?
    A) 80   B) 81   C) 82   D) 84

TEST −95

1. How many odd number less than 20?
   A) 9   B) 10   C) 11   D) 12

2. How many composite numbers less than 17?
   A) 6   B) 7   C) 8   D) 9

3. How many prime numbers less than 20?
   A) 8   B) 9   C) 10   D) 11

4. $A = 6^2$, $B = -6^2$, $C = (36 \div 6)^2$.
   Which is correct?
   A) A=C   B) A=B   C) B=C   D) B=C²

5. $A = 2$, $B = 2^{-1}$, $C = \frac{1}{2}$.
   Which s correct?
   A) A=B   B) B=C
   C) A=2B   D) A=B²

6. Find the next term of the arithmetic sequence 7,10,13,16,19.
   A) 20   B) 21   C) 22   D) 23

7. Find the next term of the arithmetic sequence 6,10,14,18,22.
   A) 24   B) 26   C) 28   D) 32

8. Find the next term of the arithmetic sequence -9,-5,-1,3,4,11.
   A) 12   B) 13   C) 14   D) 15

Find the next terms of the arithmetic sequence: -12,-9,-6,x,0,1,3,6,y

9. x=?
   A) -1   B) -2   C) -3   D) -4

10. y=?
    A) 7   B) 8   C) 9   D) 10

TEST −96

1. If 3x+4=25, what is 2x+1?
   A) 12   B) 13   C) 14   D) 15

2. If 4x+4=24, what is 3x-3?
   A) 13   B) 12   C) 11   D) 10

3. If 5x+5=75, what is 28-2x?
   A) 0   B) 7   C) 10   D) 14

4. $x \div 9 = 5$, x = ?
   A) $\frac{5}{9}$ B) $\frac{9}{5}$ C) 4   D) 45

5. $3x=3^2 \times 11$, x=?
   A) 30   B) 33   C) 34   D) 35

6. If $x \div 7 = 3$ then 3x=?
   A) 21   B) 42   C) 63   D) 84

7. If 2^5=ab, a+b=?
   A) 3   B) 4   C) 5   D) 6

8. Convert 6% to fraction.
   A) $\frac{60}{10}$ B) $\frac{60}{100}$ C) $\frac{6}{100}$ D) $\frac{6}{1000}$

Decimal	Fraction	% Percent
0.20	$\frac{20}{100}$	20%
	$\frac{3}{4}$	X
	$\frac{2}{5}$	Y

9. X=?
   A) 75%   B) 70%   C) 65%   D) 60%

10. Y=?
    A) 20%   B) 30%   C) 40%   D) 50%

TEST −97

Decimal	Fraction	% Percent
0.13		X
Y	$\frac{1}{5}$	
Z		3%

1. X=?
   A) 13%   B) 26%   C) 39%   D) 42%

2. Y=?
   A) 0.30   B) 0.03   C) 0.32   D) 0.20

3. Z
   A) 3   B) 30   C) 0.3   D) 0.03

4. Estimate: 2462×3577
   A) $8×10^6$ B) $8×10^8$
   C) $6×10$ D) $6×10^8$

5. $4\frac{2}{3} \times 5\frac{1}{7} = ?$
   A) 24   B) 26   C) 28   D) 30

6. $\frac{7}{5} - \frac{5}{7} = ?$
   A) $\frac{25}{35}$ B) $\frac{24}{30}$ C) $\frac{25}{45}$ D) $\frac{24}{35}$

7. Rounded to the tenth: 6.376
   A) 6.4   B) 6.3   C) 6.5   D) 6.48

8. Rounded to the tenth: 99.973
   A) 99   B) 99.9   C) 100   D) 101

9. Rounded to the hundredth: 64.984
   A) 64.99   B) 65.98
   C) 64.98   D) 65

10. Rounded to the hundredth: 74.897
    A) 74.90   B) 74.91
    C) 75.90   D) 74.66

TEST −98

1. Find the next term of the geometric sequence: 8,16,32,64,124.
    A) 248 B) 256 C) 267 D) 276

    Geometric sequence:
    $\dfrac{2}{3}, \dfrac{4}{6}, \dfrac{8}{12}, x, y, z$

2. x=?
    A) $\dfrac{16}{21}$ B) $\dfrac{24}{16}$ C) $\dfrac{16}{24}$ D) $\dfrac{18}{24}$

3. y=?
    A) $\dfrac{24}{38}$ B) $\dfrac{48}{32}$ C) $\dfrac{32}{38}$ D) $\dfrac{32}{48}$

4. z=?
    A) $\dfrac{64}{48}$ B) $\dfrac{64}{96}$ C) $\dfrac{96}{64}$ D) $\dfrac{86}{64}$

5. $9! \div 8! = ?$
    A) 9 B) 18 C) 36 D) 72

6. $(2! + 3!) \div 4! = ?$
    A) 3 B) $\dfrac{1}{3}$ C) 6 D) $\dfrac{1}{6}$

7. Find the reciprocal of $\dfrac{11}{7}$.
    A) $-\dfrac{11}{7}$ B) $-\dfrac{7}{11}$ C) $\dfrac{7}{11}$ D) 11

8. Find the reciprocal of $2\dfrac{1}{7}$.
    A) $\dfrac{15}{7}$ B) $-\dfrac{15}{7}$ C) $-\dfrac{7}{15}$ D) $\dfrac{7}{15}$

9. $\dfrac{1}{9} + \dfrac{9}{9} + \dfrac{10}{9} = ?$
    A) 2 B) 3.9 C) $\dfrac{20}{9}$ D) $\dfrac{90}{81}$

10. $\dfrac{11}{7} \times \dfrac{14}{44} = ?$
    A) $\dfrac{1}{2}$ B) $\dfrac{1}{3}$ C) $\dfrac{1}{7}$ D) $\dfrac{1}{4}$

TEST –99

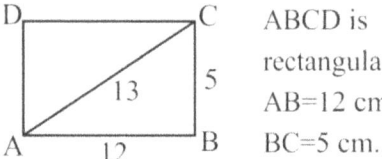

ABCD is rectangular.

1.

22 cm

24 cm

P=?
A) 46 cm   B) 48 cm
C) 90 cm   D) 92 cm

2. Find the area of ABCD rectangular.
    A) 528 cm²   B) 518 cm²
    C) 418 cm²   D) 428 cm²

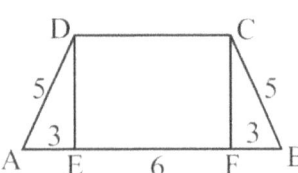

ABCD is rectangular.
AB=12 cm,
BC=5 cm.

3. Find the area of ABCD.
    A) 30   B) 34   C) 60   D) 74

4. Find the all triangle perimeter.
    A) 30   B) 45   C) 60   D) 70

ABCD is trapezoid.

5. Find the perimeter of trapezoid.
    A) 22   B) 24   C) 26   D) 28

6. Find the area of trapezoid.
    A) 30   B) 32   C) 34   D) 36

7. Find the all triangle perimeter.
    A) 12   B) 18   C) 24   D) 30

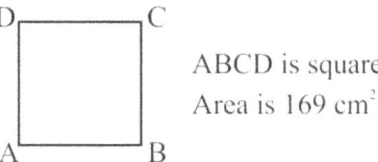

ABCD is square.
Area is 169 cm².

8. Find the one side of length.
    A) 11   B) 12   C) 13   D) 14

9. Find the perimeter.
    A) 52   B) 42   C) 39   D) 26

10.

ABCD is square.
Perimeter is 1 cm.

Find the area.

A) 1   B) $\dfrac{1}{2}$ C) $\dfrac{1}{4}$ D) $\dfrac{1}{16}$

TEST – 100

1. 21 raised to the second power.
   A) 421   B) 431   C) 441   D) 440

2. $\left(\dfrac{8}{9}\right)^{-2} = \dfrac{ab}{cd}$, $a + b + c + d = ?$

   A) 18   B) 19   C) 20   D) 21

	Jack	Marcus	Edwin
now	10	6	15
2 years ago	8	x	y
after 3 years	m	n	t

3. x=?
   A) 4   B) 5   C) 8   D) 9

4. y=?
   A) 14   B) 13   C) 12   D) 11

5. m=?
   A) 10   B) 11   C) 12   D) 13

6. n=?
   A) 9   B) 10   C) 11   D) 12

7. t=?
   A) 16   B) 17   C) 18   D) 19

8. $\dfrac{3}{4} = \dfrac{9}{x}$, $x = ?$
   A) 10   B) 11   C) 12   D) 14

9. $\dfrac{5}{7} = \dfrac{3x}{28}$, $x = ?$

   A) $\dfrac{10}{3}$ B) $\dfrac{20}{3}$ C) $\dfrac{3}{20}$ D) $\dfrac{3}{10}$

10.

   How many rectangle are at the fiqure?
   A) 3   B) 4   C) 5   D) 6

TEST – 101
(Math IQ)

1. 4, 9, 14, 19, ?
    A) 23   B) 24   C) 25   D) 25

2. 7, 13, 19, 25, x, y.,   x+y=?
    A) 68   B) 69   C) 70   D) 72

3. 4, 9, 16, 25, ?
    A) 81   B) 49   C) 36   D) 32

4. 1, 9, 25, 49, 81, x, y.,   x+y=?
    A) 180   B) 270   C) 280   D) 290

5. 1, 8, 27, 64, ?
    A) 115   B) 125   C) 135   D) 145

6. Find different number.
    A) 17   B) 19   C) 29   D) 32

7. Find different number.
    A) 38   B) 39   C) 41   D) 53

8. If 4a+4b=36, 3a+3b=?
    A) 25   B) 26   C) 27   D) 32

9. 7x+14y+21t=49, 9t+6y+3x=?
    A) 20   B) 21   C) 22   D) 24

10. UTAH→16,   KANSAS→36
    TEXAS→25,   OKLAHOMA→?
    A) 64   B) 81   C) 49   D) 36

TEST – 102
(Math IQ)

1. 642→32, 933→31, 842→42, 755→?
   A) 10   B) 13   C) 15   D) 16

2. 123→6, 723→42, 322→12, 821→?
   A) 15   B) 16   C) 18   D) 24

3. 43→12, 75→35, 84→32, 94→?
   A) 36   B) 34   C) 32   D) 30

4. 74→11.3, 93→12.6, 82→10.6, 67→?
   A) 8.13   B) 6.13   C) 13.7   D) 13.1

5. 51→26, 32→13, 34→25, 71→?
   A) 50   B) 51   C) 54   D) 60

6. UTAH, TEXAS, KANSAS, ?
   A) OKLAHOMA   B) ARKANSAS
   C) ALABAMA   D) ALASKA

7.
   34   44   ?
   A) 54   B) 45   C) 35   D) 43

8.
   9   12   ?
   A) 12   B) 18   C) 24   D) 36

9.
   0.75   0.80   ?
   A) 0.44   B) 0.45   C) 0.56   D) 0.66

10. → 8,   → ?
    → 32
    A) 12   B) 18   C) 24   D) 36

TEST – 103
(Math IQ)

1. $a^2 \square b^2$, $\rightarrow a+b$, $49 \square 144 \rightarrow$?
    A) 18   B) 19   C) 20   D) 24

2.
    9   11   ?
    A) 12   B) 13   C) 14   D) 16

3. $\left.\begin{array}{l} a + b = 11 \\ a \cdot b = 30 \end{array}\right\}$, $2a+3b=$? $(a<b)$
    A) 24   B) 25   C) 26   D) 28

4. $\left.\begin{array}{l} a + b = 12 \\ a \div b = 3 \end{array}\right\}$, $a^2+b^2=$?
    A) 90   B) 91   C) 92   D) 94

5.
    A) 24   B) 26   C) 27   D) 30

6.
    A) 28   B) 29   C) 30   D) 32

7.
8	20	12	9	7
16	4	12	15	?

    A) 10   B) 11   C) 12   D) 17

8.
16	4	6	24
3	12	8	?

    A) 1   B) 2   C) 3   D) 4

9.
74	63	42	93
28	18	8	?

    A) 27   B) 28   C) 26   D) 32

10.
64	82	79	95
5	5	8	?

    A) 6   B) 7   C) 8   D) 9

TEST – 104
(Math IQ)

1. If 3a+6b+9c=36, 21c+7a+14b=?
   A) 80   B) 81   C) 82   D) 84

2. Which numbers are divisible by 2, 3, and 7?
   A) 332   B) 333   C) 336   D) 338

rule	5074
came	1234
* sun	5674
son	809
come	829

3. sun+son=?
   A) 1638   B) 1628   C) 1248   D) 1924

4. came + come=?
   A) 10758   B) 10748
   C) 11264   D) 11374

5. lesson=?
   A) 338809   B) 348809
   C) 438809   D) 230088

6. ALABAMA □ ALASKA → 12
   UTAH □ ALASKA → 3
   KANSAS □ ARKANSAS → ?
   A) 6   B) 8   C) 9   D) 10

7. Soon, Moon, Ohio → 2, 2, 2
   Atlanta, Alaska, Utah → ?
   A) 2,2,1   B) 3,2,1   C) 3,3,1   D) 1,3,2

8.

son	game	take
can	came	start
6, 9	8, 16	?

   A) 9, 20   B) 9, 18   C) 20,12   D) 9, 12

9.

9	16	?

   A) 60   B) 64   C) 49   D) 81

10. A) 25   B) 75
    C) 100   D) 125

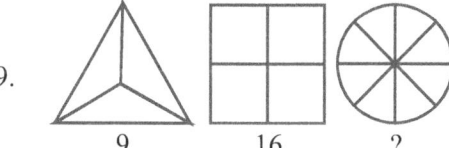

TEST – 105
(Math IQ)

* **In questions 1 and 2, the numbers in group II stand for the words in group I, when each letter has been coded with a specific numeral. Find the number which corresponds to the word indicated by the question mark.**

1. Alba $\left.\begin{array}{l}\\ \text{Eden}\\ \text{van}\\ \text{man}\end{array}\right\}$ $\left\{\begin{array}{l}716\\1231\\616\\4546\end{array}\right.$, Alabama=?

    A) 1213131   B) 1213171
    C) 1213141   D) 1214151

2. man $\left.\begin{array}{l}\\ \text{old}\\ \text{new}\\ \text{van}\end{array}\right\}$ $\left\{\begin{array}{l}923\\123\\378\\456\end{array}\right.$, old+new=?

    A) 379   B) 834   C) 479   D) 579

3. Alba□Alma→4, Eden□Edna→3,
    Ennis□Edna→3, Alabama□Dallas→?

    A) 5   B) 6   C) 7   D) 8

4. Dallas□Alba→4,   Alba□Alma→4,
    Waco□Van→2, Son□Soon→?

    A) 2   B) 3   C) 4   D) 5

5. Came□Alba→3,   Alma□Van→3,
    Cool□Come→3, Cooper□Cool→?

    A) 3   B) 4   C) 5   D) 6

* Dallas $\left.\begin{array}{l}\\ \text{Denton}\\ \text{Denver}\\ \text{Re no}\\ \text{Toledo}\end{array}\right\}$ $\left\{\begin{array}{l}783518\\123324\\0568\\156786\\156950\end{array}\right.$

6. Toledo→?

    A) 156950   B) 123324
    C) 783518   D) 156786

7. Dallas→?

    A) 123324   B) 783518
    C) 156786   D) 156950

8. Reno+Dallas→?

    A) 123982   B) 123972
    C) 123992   D) 123892

9. Reno+Toledo→?

    A) 784082   B) 784084
    C) 784085   D) 784086

10. Toll+ton→?

    A) 8019   B) 8629
    C) 8719   D) 8819

TEST – 106
(Math IQ)

Which of the below should be replaced in the question mark (?)

1.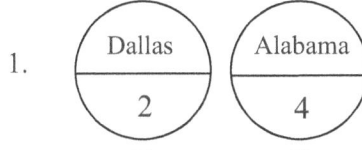

   A) 1   B) 2   C) 3   D) 4

2.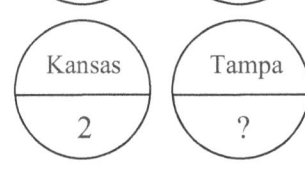

   A) 16   B) 26   C) 36   D) 46

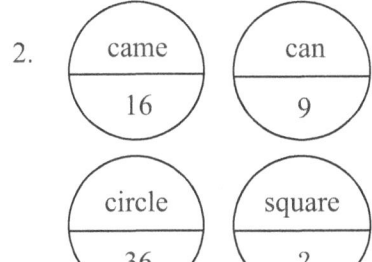

3. x=?
   A) 5   B) 4   C) 3   D) 2

4. y=?
   A) 8   B) 9   C) 10   D) 12

5. m+n=?
   A) 18   B) 20   C) 22   D) 24

6. (x+n)²=?
   A) 121   B) 144   C) 225
   D) 324

*

16	14	12	x	N
12	10	8	y	K
72	50	M	z	L

7. x+y=?
   A) 13   B) 15   C) 18   D) 16

8. m+z=?
   A) 44   B) 46   C) 48   D) 50

9. l − n=?
   A) 0   B) 2   C) 4   D) 6

10. m+z+l=?
    A) 48   B) 52.5   C) 68   D) 78

TEST – 107
(Math IQ)

* **How many parallelograms are in each figure?**

1. A) 5   B) 6
   C) 7   D) 8

2. A) 10   B) 11
   C) 12   D) 13

3. A) 7   B) 8
   C) 9   D) 10

4. A) 8   B) 9
   C) 10   D) 12

* **How many rectangles are in each figure?**

5. A) 11   B) 12
   C) 13   D) 14

6. A) 3   B) 4
   C) 5   D) 6

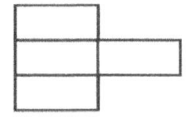

7. A) 6   B) 7
   C) 8   D) 9

8. A) 9   B) 10
   C) 11   D) 12

9. A) 10   B) 11
   C) 12   D) 13

10. A) 12   B) 15
   C) 16   D) 18

TEST – 108
(Math IQ)

* A: 64, 32, 16, 8, x
   B: 24, 6, 6, 8, y
   C: 12, 3, 3, 4, z

In accordance with the relationship figure. Find the number which corresponds to the place indicated by the question mark.

1. x=?
    A) 7   B) 6   C) 5   D) 4

6. 6□=12, 7◻28, 8◻24, ((8□)◻)◻=?
    A) 162   B) 172   C) 182   D) 192

2. y=?
    A) 4   B) 5   C) 6   D) 7

7. 9⊔=3, 8∏=16, 4⊐=2, (((27⊔)∏)⊐)=?
    A) 10   B) 9   C) 12   D) 16

3. z=?
    A) 1   B) 2   C) 3   D) 4

8. 2⊖=4, 3⊖=9, 4⊕=2, 6⊕=3,
    ((10⊖)⊕)=?
    A) 60   B) 50   C) 45   D) 40

* A: 84, 64, 76, 92, 74
   B: 32, 24, 42, 18, x
   C: 6, 8, 8, 8, y

4. x=?
    A) 28   B) 24   C) 20   D) 18

9. 12⊟=4, 24⊟=8, 6▯=18, 7▯=21,
    ((2▯+9⊟)⊟)=?
    A) 3   B) 6   C) 9   D) 12

5. y=?
    A) 8   B) 9   C) 10   D) 16

10. 2■=8, 1■=1, 3■=27, (4■+5■)=?
    A) 180   B) 184   C) 189   D) 190

TEST – 109
(Math IQ)

Which number should be replaced in the question mark (?)?

1.

0.50   0.25   0.17   ?

A) 0.16   B) 0.15   C) 0.14   D) 0.13

2.

1/4   1/9   ?

A) 1/4   B) 1/8   C) 1/16   D) 1/64

3.

1/64   1/27   ?

A) 1/216   B) 1/108   C) 1/36   D) 1/98

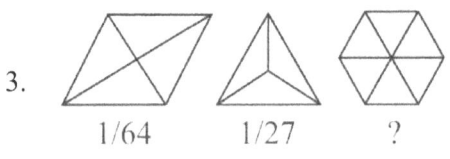

*

*	∟	V	N
⊔	⌂	⌂	⬡
>	□	M	n
|	x	Y	z

4.   m=?

A) ∟   B) ∧   C) Δ   D) □

5.   n=?

A) □   B) ⌂   C) ○   D) Δ

6.   x=?

A) ∧   B) Δ   C) □   D) ⌂

7.   y=?

A) Δ   B) □   C) ⌂   D) ○

8.   z=?

A) Δ   B) □   C) ⊐   D) ⊔

*

*	|	V	T
⊔	Δ	□	⌂
∟	Δ	M	n

9.   m=?

A) Δ   B) □   C) ⊔   D) ⌂

10. n=?

A) ⊓   B) Δ   C) □   D) ⌂

TEST – 110
(Math IQ)

1. What is the sum of all the even number strictly between 1 and 9?

   A) 20   B) 22   C) 24   D)26

2. What is the sum of all the odd number strictly between 2 and 10?

   A) 21   B) 22   C) 23   D)30

3. AA is the two digit smallest odd number A?

   A) 5   B) 4   C) 3   D) 1

4. BB is two digit biggest positive odd number (B+B)?

   A) 20   B) 18   C) 16   D)18

5. AB is two digit biggest prime number . (A+B)?

   A) 16   B) 14   C) 13   D) 17

6. $\triangle+\triangle=6$, $\triangle+\triangle=8$, $2\triangle+\square=?$

   A)11   B) 12   C) 14   D)16

7. $\triangle-\triangle=2$, $\square-\triangle=?$

   A) 1   B) 2   C) 3   D) 4

8. $\square\times\triangle=12$, $\square\times\triangle=20$, $\triangle\times\triangle=?$

   A) 12   B) 15   C) 16   D)18

9. CAN, MAN ← 123, 423 IF C+M?

   A) 7   B) 6   C) 5   D) 4

10. SUN, SON ← 987, 967 IF U+O?

    A) 11   B) 12   C) 13   D) 14

TEST – 111
(Math IQ)

1. Which number below is the smallest prime number?

   A) 7   B) 5   C) 3   D) 2

2. Which is the two digit smallest prime number?

   A) 19   B) 17   C) 13   D) 11

3. What is the sum one digit prime numbers?

   A) 18   B) 17   C) 16   D) 14

4. What is the sum prime number between 10 and 15?

   A) 24   B) 23   C) 22   D) 21

5. Which is the biggest two digit prime number?

   A) 98   B) 97   C) 96   D) 94

6. What is the next prime number that comes after 31?

   A) 36   B) 37   C) 38   D) 41

7. What is the next prime number that comes after 41?

   A) 42   B) 43   C) 47   D) 49

8. How many prime numbers are these less than 14?

   A) 6   B) 7   C) 8   D) 9

9. HOME, SAME ← 1234, 5634 if seem?

   A) 5443   B) 6446
   C) 6331   D) 3114

10. COME, CAME ←1234, 1534
    IF (O+A)=?

    A) 10   B) 9   C) 8   D) 7

TEST – 112
(Math IQ)

In questions 1 and 10 the numbers in group II stand for the words in group I, when each letter has been coded with a specific numeral. Find the number which corresponds to the word indicated by the question mark.

1. BAKE, CAKE → 1234, 5234
   IF (B+C)=?
   A) 6   B) 5   C) 4   D) 3

2. BALL, BAND → 1233, 1245
   IF (B+A+N)=?
   A) 10   B) 9   C) 7   D) 6

3. BAR, CAR, CAN → 426, 423, 123
   IF (B+N)?
   A) 8   B) 7   C) 6   D) 5

4. BAY, BID, BIT → 146, 123, 145
   IF (I+D+Y)?
   A) 13   B) 12   C) 9   D) 11

5. IF CAR, BUT, BUS → 457, 123, 456
   THEN (C+R+S)?
   A) 8   B) 9   C) 10   D) 17

6. DAB, CAB, DAM → 125, 423, 123
   IF (D+A+M)?
   A) 10   B) 8   C) 7   D) 6

7. DOOL, POOL, DOCK →
   1223,4223,1256 IF COOK?
   A) 6225   B) 3226   C) 5226   D) 4226

8. MAN, HEN → 123, 453 IF MEN ?
   A) 123   B) 143   C) 153   D) 351

9. EGG, EVE, FIT → 456, 131, 122
   IF (F+V+G)?
   A) 12   B) 11   C) 10   D) 9

10. CAME, GAME, SAME→1234, 5234,
    6234 IF (C+G+S)?
    A) 12   B) 11   C) 10   D) 9

TEST – 113
(Math IQ)

1. What is the sum of all the odd numbers strictly between 2 and 8 ?
   A) 15   B) 20   C) 22   D) 23

2. What is the sum of all the even numbers strictly between 5 and 13?
   A) 24   B) 36   C) 28   D) 30

3. Which number below is the largest two digit even number?
   A) 97   B) 98   C) 99   D) 96

4. What is 1/9 of the largest two digit positive number?
   A) 32   B) 11   C) 34   D) 3

5. The sum of two consecutive odd integers is equal to 8. What is the product of the two integers?
   A) 14   B) 17   C) 15   D) 12

6. The product of two consecutive odd integers is equal to 15. What is the sum of the two integers?
   A) 8   B) 10   C) 11   D) 12

7. What is the product of all the odd numbers strictly between 3 and 9?
   A) 945   B) 816   C) 718   D) 619

8. What is the next prime number that comes after 11?
   A) 15   B) 14   C) 12   D) 13

9. How many odd numbers are there less than 12?
   A) 6   B) 7   C) 8   D) 19

10. If the father is 32 years old and the son is 4 what will be the sum of their ages in six years?
    A) 54   B) 48   C) 46   D) 60

TEST – 114
(Math IQ)

1. UTAH → TEXAS → OREGON → ?

    A) NEVADA   B) MONTANA
    C) OKLAHOMA   D) MINNESOTA

2. OHIO ↔ UTAH, TEXAS ↔ MAINE,
    ALASKA ↔ ?

    A) KENTUCKY   B) KANSAS
    C) ARKANSAS   D) INDIANA

3. ALABAMA, ALASKA, ARIZONA,?

    A) CALIFORNIA   B) ARKANSAS
    C) COLORADO   D) FLORIDA

4. MONTANA, MISSOURI,
    MISSISSIPPI, ?

    A) MINNESOTA   B) KANSAS
    C) IOWA   D) INDIANA

5. ALABAMA → 4, ALASKA → 3
    ARKANSAS → 3, OKLAHOMA → ?

    A) 1   B) 2   C) 3   D) 4

6. COLORADO → 3, OHIO → 2
    OREGON → 2, OKLAHOMA → ?

    A) 1   B) 2   C) 3   D) 4

7. UTAHΔTEXAS → 20,
    OREGONΔOHIO → 24,
    NEVADAΔFLORIDA → ?

    A) 40   B) 42   C) 43   D) 44

8. ALABAMA Δ KANSAS → 1
    MONTANA Δ UTAH → 3
    OKLAHOMA Δ NEVADA → ?

    A) 2   B) 3   C) 4   D) 5

9. OREGON Δ OHIO → 2,
    NEBRASKA Δ OHIO → 4,
    MINNESOTA Δ OHIO → ?

    A) 3   B) 4   C) 5   D) 6

10. ALABAMA Δ ALASKA → 7,
    ALABAMA Δ ARIZONA → 6,
    ALABAMA Δ ARKANSAS →?

    A) 7   B) 8   C) 9   D) 10

TEST – 115
(Math IQ)

The figures in group II stand for the words in group I, when each letter has been coded with a specific numeral. Find the number which corresponds to the word indicated by the question mark.

1. MEN, MAN, HEN, CAN →
   ⊓⊔⊕, ◻⋒⊕, ◻⊔⊕, ○⊔◻ IF HC?
   A) ⊓⊔ B) ◻⋒ C) ◻⊔ D) ⊓○

2. BUY, MEN, MAN, PAY →
   ◻⊔○, ◻◻○, ⋒◻⊕, ◿⊎⊕ IF PAY?
   A) ◻⊔○   B) ◻◻○
   C) ⋒◻⊕   D) ◿⊎⊕

3. ADEL, ALTA, ADAK, AMES →
   ⊕⊎⊕◿, ⊕⊎◻◻, ⊕◻○⊕, ⊕⊔◻⊓ IF ALTA?
   A) ⊕⊎⊕◿   B) ⊕⊎◻◻
   C) ⊕◻○⊕   D) ⊕⊔◻⊓

4. TULSA, ALTON → ⊓◻⊔⊕○,
   ○⊔⊓⊎◿ IF SOON?
   A) ◿⊎⊓⊔   B) ⊕⊎⊎◿
   C) ⊕⊔⊓○   D) ◻◻⋒⊔

5. LAMAR, LARGO →
   ◻⊔◻⊔⋒, ◻⊔⋒◻◿⊎ IF MOOR?
   A) ⊎◿⋒⊔   B) ⋒◿◿⊔
   C) ◻⊎⊎⋒   D) ⊎◻◻⋒

6. WACO, WALL → ◻⋒⊎⊔, ◻⋒◿◻
   IF CALL?
   A) ◿⋒⋒◿   B) ◻◻◿⋒
   C) ⊔◿◻⊔   D) ⊎⋒◿◿

7. ETNA→◿⋒⊎⊔ IF TEN?
   A) ⋒◿⊎   B) ⋒⊎◿
   C) ◻◻⊎   D) ⊎◻◿

8. TULSA, BELEN → ◿◻⊎⋒⊔,
   ○◻◻◻⋒ IF ELSE?
   A) ⊓◻◻◻   B) ◻⊎⋒◻
   C) ◻⊓⋒◻   D) ⋒⊓⊎◻

9. TOK, CAN → ◻⊎⊓, ◻○⊔ IF COOK?
   A) ⊔○⊔⊔   B) ○◻◻⊔
   C) ○⊓⊔◻   D) ◻⊎⊎⊓

10. MERNA → ⊓⊎◻⊔◿ IF NEAR?
    A) ◿⊔◻⊎   B) ◻⊔⊎◿
    C) ⊔⊎◿◻   D) ◿◻⊎⊔

PROBLEM – 116

1. 3030-30-3=

2. 144-44+4=

3. 248+48-38=

4. 999-888+111=

5. 624-24-48=

6. 3040-340-40=

7. 124+38+14=

8. 1999-999+99=

9. 2030-330-30=

10. 777-444-44=

11. 884-84+48=

12. 6060-660-60=

13. 19999-18888-18=

14. 555-505-5=

15. 299-99-9=

16. 766-606-6=

17. 99-88-8=

18. 124-24-4=

19. 827-72-77=

20. 4040-40-4=

PROBLEM – 117

1. 364+724=

2. 984+87+17=

3. 1999+1888+77=

4. 327+484=

5. 927+189=

6. 877+366=

7. 999+888+777=

8. 148 +824 =

9. 674+748=

10. 1699+3894=

11. 148+324+180=

12. 1999+ 1888+1777=

13. 199+288+377 =

14. 555+444+222=

15. 777+444+111=

16. 2002+2020+2200=

17. 701-107+207=

18. 824+724+624=

19. 1984+1977+1877=

20. 19214+18412+1777=

PROBLEM – 118

1. 88 ÷ 4=

2. 99 ÷ 3=

3. 108 ÷ 9=

4. 78 ÷ 13 =

5. 777 ÷ 3=

6. 972 ÷ 6=

7. 2271÷ 3=

8. 996 ÷ 4=

9. 288 ÷ 12=

10. 824 ÷ 4=

11. 84 ÷ 21=

12. 96 ÷ 3=

13. 256 ÷ 18=

14. 360 ÷ 40=

15. 720 ÷ 24=

16. 960 ÷ 30=

17. 976 ÷ 6=

18. 879 ÷ 6=

19. 1992 ÷ 6=

20. 484 ÷ 4=

PROBLEM – 119

1. The remainder of 95 ÷ 5 is ………

2. The remainder of 76 ÷ 5 is ………

3. The remainder of 145 ÷ 12 is ………

4. The remainder of 245 ÷ 3 is ………

5. The remainder of 289 ÷ 24 is ………

6. The remainder of 327 ÷ 5 is ………

7. The remainder of 677 ÷ 4 is ………

8. The remainder of 777 ÷ 3 is ………

9. The remainder of 984 ÷ 4 is ………

10. The remainder of 827 ÷ 5 is ………

11. The remainder of $(3^2+4^2)÷4$ is ……

12. The remainder of $(6^2+6) ÷ 3$ is ……

13. The remainder of $(3^2+4^2) ÷ 5$ is ……

14. The remainder (999-99) ÷ 9 is …..

15. The remainder of 97 ÷ 5 is …..

16. The remainder of 878 ÷ 15 is …..

17. The remainder of 1484 ÷ 5 is …..

18. The remainder of 977 ÷15 is …..

19. The remainder of 506÷ 22 is …..

20. The remainder of 1972÷ 12 is …..

PROBLEM – 120

1. The product of 6 and 26 is ...

2. The product of 68 and 24 is ...

3. The product of 9 and 19 is ...

4. The product of 3^2 and 2^3 is ...

5. The product of 4^2 and 2^4 is ...

6. The product of $\left(\dfrac{1}{2}\right)^2$ and $\left(\dfrac{2}{1}\right)^2$ is ...

7. The product of 8 and 18 is ...

8. The product of 2 and 22 is ...

9. The product of 1^{22} and 22^1 is ...

10. The product of 5^2 and 2^5 is ...

11. The product of 6^2 and 3^2 is ...

12. The product of 3^3 and 2^2 is ...

13. The product of 3^4 and 4^3 is ...

14. The product of $\left(\dfrac{3}{2}\right)^3$ and $\left(\dfrac{2}{3}\right)^1$ is ...

15. The product of 5 and 15 is ...

16. The product of 2^4 and 1^4 is ...

17. The product of 10^4 and $\dfrac{1}{10^4}$ is ...

18. The product of 2^6 and 2^{-6} is ...

19. The product of 12 and 21 is ...

20. The product of 10^{10} and 10^{-10} is ...

PROBLEM – 121

1. $10 \times 27 =$

2. $100 \times 2.4 =$

3. $10 \times 3.9 =$

4. $100 \times 37 =$

5. $10 \times 2.9 =$

6. $10 \times 28 =$

7. $100 \times 1.23 =$

8. $1000 \times 0.321 =$

9. $10^2 \times 0.27 =$

10. $10^3 \times 0.674 =$

11. $10 \times 2^4 =$

12. $5^4 \times 10 =$

13. $5^3 \times 100 =$

14. $\left(\dfrac{1}{2}\right)^2 \times 100 =$

15. $\left(\dfrac{1}{3}\right)^3 \times 2700 =$

16. $6^2 \times 10^2 =$

17. $8^2 \times 10^3 =$

18. $2^3 \times 10^3 =$

19. $5^4 \times 10^4 =$

20. $6^3 \times 10^3 =$

PROBLEM – 122

1. $(9 \times 4 - 4) =$

2. $(7 \times 9 - 9) =$

3. $12 \times 13 - 11 =$

4. $3^2 \times 4 - 3 =$

5. $10^2 \times 5 - 4 =$

6. $2^3 \times 3^2 - 4 =$

7. $17 \times 7 - 6 =$

8. $19 \times 9 - 8 =$

9. $10^2 \times 7 - 17 =$

10. $10^3 \times 3 - 300 =$

11. $9 \times 5 - 4 =$

12. $9^2 \times 3 - 14 =$

13. $3^4 \times \dfrac{1}{3} - 3 =$

14. $5^3 \times 2^3 - 4^2 =$

15. $17 \times 7 - 6 =$

16. $27 \times 7 - 17 =$

17. $8 \times 4 - 3 \times 7 =$

18. $12 \times 13 - 4 \times 9 =$

19. $10^4 \times 12 - 10^2 \times 13 =$

20. $10^3 \times 7 - 19 \times 100 =$

PROBLEM – 123

1. Which digit is in the ten place in 487?

2. Which digit is in the ten place in 1246?

3. Which digit is in the ten place in 9867?

4. In 197643, which digit is in the thousands place?

5. In 29814, which digit is in the thousands place ?

6. In 9977632, which digit is in the thousands place ?

7. The hundreds digit of 6984 is …

8. The hundreds digit of 12345 is …

9. The hundreds digit of 98726 is ….

10. The hundreds digit of 19873 is ….

11. 72+44+77=

12. 84+3+7+8=

13. 99+88+67=

14. 9+19+29+39=

15. 8+18+18+86=

16. 17+27+33+44+55=

17. 99+88+77=

18. 36+46+56+76=

19. 36+46+76=

20. 9+19+8+18=

PROBLEM – 124

1. Round off 327 to the nearest ten

2. Round off 948 to the nearest ten

3. Round off 321 to the nearest ten

4. Round off 652 to the nearest ten

5. Round off 765 to the nearest ten

6. Round off 9843 to the nearest hundred

7. Round off 12375 to the nearest hundred

8. Round off 9872 to the nearest hundred

9. Round off 5643 to the nearest hundred

10. Round off 1234 to the nearest hundred

11. Round off 0.4872 to the nearest hundredth

12. Round off 0.1234 to the nearest hundredth

13. Round off 0.9872 to the nearest hundredth

14. Round off 0.8267 to the nearest tenth

15. Round off 0.4276 to the nearest tenth

16. 3284 rounded to the nearest ten is ….

17. 6542 rounded to the nearest hundred is…

18. 22×33=

19. 44×11=

20. $3^4 \times 12$=

PROBLEM – 125

1. $17^2 =$

2. $19^2 =$

3. $21^2 =$

4. $\left(\dfrac{5}{7}\right)^2 =$

5. $(32)^2 =$

6. $\left(\dfrac{2}{3}\right)^4 =$

7. $\left(\dfrac{7}{5}\right)^3 =$

8. $15^2+14^2 =$

9. $13^2+12^2 =$

10. $9^2+8^2+7^2 =$

11. $1^2+2^2+3^2+4^2 =$

12. $1^3+2^3+3^3 =$

13. $9^2-8^2 =$

14. $10^2-9^2 =$

15. $12^2-2^2 =$

16. $19^2-9^2 =$

17. $21^2-20^2 =$

18. $33^2-22^2 =$

19. $9^2-8^2-7^2+10^2 =$

20. $20^2-10^2-9^2 =$

PROBLEM – 126

1. 808-406=

2. 324-342=

3. 328-114=

4. 199-88=

5. 984-127=

6. 55×24=

7. 36×12=

8. 67×16=

9. 88×22=

10. 70×14=

11. 684-127=

12. 75×14=

13. 95×12=

14. 105×17=

15. 64×15=

16. 94×14=

17. 84×13=

18. 27×19=

19. 32×44=

20. 27×12=

PROBLEM – 127

1. $16 \times 14 \div 7 =$

2. $25 \times 10 \div 2 =$

3. $72 \times 18 \div 6 =$

4. $18 \times 3 \div 2 =$

5. $36 \times 6 \div 6 =$

6. $96 \times 6 \div 2 =$

7. $125 \times 25 \div 5 =$

8. $10^2 \times 10 \div 5 =$

9. $8^2 \times 16 \div 8 =$

10. $9^2 \times 3 \div 3 =$

11. $48 \times 12 \div 3 =$

12. $144 \times 9 \div 12 =$

13. $196 \times 7 \div 14 =$

14. $288 \times 4 \div 12 =$

15. $14^2 \times 4 \div 12 =$

16. $169 \times 3 \div 13 =$

17. $64 \times 5 \div 5 =$

18. $77 \times 7 \div 14 =$

19. $88 \times 8 \div 4 =$

20. $10^2 \times 12 - 19 \times 12 =$

PROBLEM – 128

Find the greatest common factor of the following.

1. 6 and 12:

2. 5 and 10:

3. 7 and 21:

4. 15 and 25:

5. 64 and 16:

Find the least common multiple of the following.

6. 4 and 3:

7. 5 and 2:

8. 7 and 3:

9. 2,4 and 3 :

10. 3,4 and 7:

Convert to Decimal

11. $\dfrac{7}{3}$

12. $\dfrac{9}{7}$

13. $\dfrac{27}{5}$

14. $\dfrac{37}{7}$

15. $\dfrac{97}{9}$

16. $\dfrac{77}{24}$

17. $\dfrac{127}{25}$

18. $\dfrac{145}{12}$

19. $\dfrac{78}{5}$

20. $\dfrac{98}{9}$

PROBLEM – 129

Compare fraction: $(>, <, =, \geq, \leq)$

1. $\dfrac{7}{11}$ and $\dfrac{6}{10}$

2. $\dfrac{3}{2}$ and $\dfrac{4}{3}$

3. $\dfrac{9}{7}$ and $\dfrac{11}{6}$

4. $\dfrac{9}{14}$ and $\dfrac{7}{12}$

5. $\dfrac{12}{11}$ and $\dfrac{11}{12}$

6. $\dfrac{4}{3}$ and $\dfrac{5}{4}$

7. $\dfrac{7}{11}$ and $\dfrac{8}{10}$

8. $\dfrac{4}{9}$ and $\dfrac{5}{8}$

9. $\dfrac{7}{5}$ and $\dfrac{3}{4}$

10. $\dfrac{1}{7}$ and $\dfrac{1}{6}$

11. $32^2 =$

12. 44^2

13. $55^2 =$

14. $77^2 =$

15. $21^2 + 11^2 =$

16. $28^2 - 18^2 =$

17. $11^2 + 12^2 + 13^2 =$

18. $1^2 + 2^2 + 3^2 + 4^2 =$

19. $(1)^9 + (9)^1 =$

20. $(1)^{99} + (99)^1 =$

PROBLEM – 130

1. $444 \div 22 =$

2. $777 \div 21 =$

3. $684 \div 4 =$

4. $1296 \div 12 =$

5. $1545 \div 15 =$

6. $1975 \div 15 =$

7. $1670 \div 10 =$

8. $1978 \div 12 =$

9. $9988 \div 22 =$

10. $6976 \div 12 =$

11. $(3^2 + 4^2) \div 5 =$

12. $(6^2 + 8^2) \div 10 =$

13. $(5^2 + 10^2) \div 5 =$

14. $(7^2 + 7) \div 7 =$

15. $(12^2 + 6^2) \div 6 =$

16. $(8^2 + 7^2) \div 10 =$

17. $(1^2 + 2^2 + 3^2) \div 14 =$

18. $(10^2 + 8^2) \div 16 =$

19. $(20^2 - 10^2) \div 3 =$

20. $(10-2)^2 \div 16 =$

PROBLEM – 131

1. $\dfrac{5}{6} - \dfrac{1}{5} =$

2. $\dfrac{11}{12} - \dfrac{2}{12} =$

3. $\dfrac{24}{27} - \dfrac{3}{27} =$

4. $\dfrac{4}{5} - \dfrac{1}{2} =$

5. $\dfrac{6}{5} - \dfrac{1}{5} =$

6. $\dfrac{4}{9} - \dfrac{1}{5} =$

7. $\dfrac{7}{8} - \dfrac{6}{5} =$

8. $\dfrac{4}{5} - \dfrac{1}{2} =$

9. $\dfrac{11}{12} + \dfrac{1}{3} =$

10. $\dfrac{2}{7} + \dfrac{1}{4} =$

11. $\dfrac{7}{3} + \dfrac{1}{4} =$

12. $\dfrac{9}{8} + \dfrac{1}{4} =$

13. $\dfrac{4}{3} + \dfrac{1}{7} =$

14. $\dfrac{6}{5} + \dfrac{4}{8} =$

15. $\dfrac{7}{2} + \dfrac{1}{7} =$

16. $\dfrac{9}{10} + \dfrac{4}{5} =$

17. $\dfrac{6}{9} + \dfrac{1}{8} =$

18. $\dfrac{12}{11} + \dfrac{13}{14} =$

19. $\dfrac{2}{3} + \dfrac{1}{7} =$

20. $\dfrac{4}{7} + \dfrac{1}{3} =$

PROBLEM – 132

1. Two days are …… minute.

2. Six days are …… hour.

3. Two weeks are …… hour.

4. 3 hours are …… minute.

5. Half day is …… minute.

6. 98 days are …… week.

7. 5 weeks are …… day.

8. 4 years are …… day.

9. 105 days are …… week.

10. 104 weeks are …… year.

11. 3 years are …… week.

12. 64 weeks are …… day.

13. 2 weeks are …… hour.

14. How many prime number less than 20?

15. How many prime number less than 32?

16. How many composite number less than 61?

17. How many composite number less than 99?

18. How many prime number are between 30 and 60?

19. How many prime number are between 40 and 60?

20. Add least prime and composite number.

PROBLEM − 133

1. The LCM of 12 and 24 is:

2. The LCM of 14 and 6 is:

3. The LCM of 12 and 15 is:

4. The LCM of 9 and 12 is:

5. The LCM of 17 and 12 is:

6. The LCM of 16 and 24 is:

7. The LCM of 15 and 20 is:

8. The LCM of 25 and 35 is:

9. The LCM of 30 and 45 is:

10. The LCM of 9 and 16 is:

11. $2.4 \times 10^2 =$

12. $26 \times 10^{-1} =$

13. $36 \times 10^{-2} =$

14. $123 \times 10^{-2} =$

15. $6^2 \times 10^{-1} =$

16. $18^2 \times 10^{-2} =$

17. $1.24 \times 10^{-2} =$

18. $(1.2)^2 \times 10^{-3} =$

19. $(0.8)^2 \times 10^{-1} =$

20. $(0.3)^2 \times 10^2 =$

PROBLEM – 134

Reduce the following fraction.

1. $\dfrac{6}{9}$

2. $\dfrac{15}{25}$

3. $\dfrac{35}{25}$

4. $\dfrac{12}{16}$

5. $\dfrac{42}{24}$

6. $\dfrac{48}{56}$

7. $\dfrac{72}{96}$

8. $\dfrac{36}{27}$

9. $\dfrac{32}{48}$

10. $\dfrac{24}{27}$

Adding mixed numbers.

11. $2\dfrac{1}{2} + 3\dfrac{2}{3}$

12. $5\dfrac{2}{3} + 2\dfrac{1}{7}$

13. $2\dfrac{1}{6} + 3\dfrac{1}{5}$

14. $4\dfrac{1}{7} + 3\dfrac{2}{3}$

15. $3\dfrac{1}{2} + 5\dfrac{2}{3}$

16. $2\dfrac{1}{3} + 3\dfrac{1}{4}$

17. $3\dfrac{1}{4} + 4\dfrac{1}{3}$

18. $9\dfrac{1}{3} + 8\dfrac{1}{4}$

19. $2\dfrac{1}{3} + 4\dfrac{1}{4}$

20. $3\dfrac{1}{7} + 3\dfrac{1}{6}$

PROBLEM − 135

Dividing fractions.

1. $\dfrac{6}{5} \div \dfrac{1}{4}$

2. $\dfrac{11}{12} \div \dfrac{9}{10}$

3. $\dfrac{14}{15} \div \dfrac{2}{3}$

4. $6 \div \dfrac{1}{8}$

5. $\dfrac{1}{12} \div 6$

6. $\dfrac{4}{3} \div \dfrac{1}{7}$

7. $\dfrac{6}{9} \div \dfrac{1}{6}$

8. $\dfrac{2}{3} \div \dfrac{1}{4}$

9. $\dfrac{2}{3} \div \dfrac{1}{7}$

10. $\dfrac{6}{9} \div \dfrac{1}{3}$

Dividing complex fractions.

11. $\dfrac{\frac{1}{2}}{\frac{3}{4}}$

12. $\dfrac{\frac{6}{9}}{\frac{1}{3}}$

13. $\dfrac{\frac{7}{2}}{\frac{1}{4}}$

14. $\dfrac{\frac{2}{9}}{\frac{4}{7}}$

15. $\dfrac{\frac{12}{11}}{\frac{2}{3}}$

16. $\dfrac{\frac{15}{7}}{\frac{1}{3}}$

17. $\dfrac{\frac{12}{15}}{\frac{11}{13}}$

18. $\dfrac{\frac{6}{9}}{\frac{1}{7}}$

19. $\dfrac{\frac{2}{5}}{\frac{1}{3}}$

20. $\dfrac{\frac{1}{7}}{\frac{1}{3}}$

PROBLEM − 136

Dividing mixed numbers.

1. $2\frac{1}{3} \div 4\frac{1}{3}$

2. $6\frac{1}{2} \div 4\frac{1}{2}$

3. $3\frac{1}{4} \div 2\frac{1}{3}$

4. $9\frac{1}{2} \div 9\frac{1}{2}$

5. $2\frac{1}{3} \div 3\frac{1}{4}$

6. $4\frac{1}{2} \div 2\frac{1}{5}$

7. $6\frac{1}{2} \div 4\frac{1}{7}$

8. $2\frac{1}{3} \div 2\frac{1}{5}$

9. $3\frac{1}{3} \div 2\frac{1}{2}$

10. $2\frac{1}{2} \div 3\frac{1}{2}$

Changing decimal to fraction.

11. 0.7 =

12. 0.9 =

13. 0.12 =

14. 0.64 =

15. 0.123 =

16. 0.71 =

17. 0.640 =

18. 0.78 =

19. 0.98 =

20. 0.424 =

PROBLEM – 137

Changing fraction to decimal.

1. $\dfrac{1}{4}$

2. $\dfrac{3}{4}$

3. $\dfrac{7}{20}$

4. $\dfrac{17}{25}$

5. $\dfrac{19}{25}$

Changing decimal to percent.

6. 0.72 :

7. 0.64 :

8. 0.09 =

9. 0.6 :

10. 0.84

Changing percent to decimals.

11. 24%=

12. 36%=

13. 80%=

14. 96%=

15. 12% =

Changing fraction to percent.

16. $\dfrac{3}{8}$

17. $\dfrac{3}{4}$

18. $\dfrac{9}{20}$

19. $\dfrac{5}{26}$

20. $\dfrac{7}{3}$

PROBLEM – 138

Finding percent of a number.

1. What is 20 % of 60?

2. What is 30 % of 40 ?

3. What is 24 % of 50?

4. What is 27 % of 60?

5. What is 12 % of 300?

6. 30 is what percent of 90?

7. 40 is what percent of 120?

8. 15 is what percent of 75?

9. 9 is what percent of 108?

10. 16 is what percent of 144?

11. Find the percent decrease from 60 to 40

12. Find the percent decrease from 100 to 80

13. Find the percent decrease from 200 to 160

14. Find the percent decrease from 120 to 80

15. 60% of what number is 120

16. 30% of what number is 90

17. What number is 35% of 140

18. Find the percent increase from 80 to 100

19. Find the percent increase from 120 to 160

20. Find the percent increase from 180 to 200

PROBLEM – 139

Multiplication in scientific notation.

1. $(2\times10^2) \times (4\times10^4)$

2. $(3\times10^3) \times (5\times10^4)$

3. $(7\times10^6) \times (3\times10^7)$

4. $(5.2\times10^3) \times (4\times10^3)$

5. $(7\times10^4) \times (3\times10^5)$

Division in scientific notation.

6. $(9\times10^4) \div (3\times10^2)$

7. $(12\times10^6) \div (6\times10^3)$

8. $(24\times10^6) \div (8\times10^2)$

9. $(4.24\times10^8) \div (4\times10^2)$

10. $(48\times10^4) \div (8\times10^3)$

11. The 6th term in the arithmetic sequence 4,9,14,19,24,…… is

12. The 7th term in the arithmetic sequence 7,13,19,25,31,…… is

13. The 8th term in the arithmetic sequence 8,14,20,26,32,…… is

14. The 9th term in the arithmetic sequence 9,16,23,30,37…… is

15. The 10th term in the arithmetic sequence 8,4,0,-4,-8,-12…… is

16. $\sqrt{1296}$: ……

17. $\sqrt{1225}$: ……

18. $\sqrt{1681}$: ……

19. $\sqrt{2601}$: ……

20. $\sqrt{3721}$: ……

PROBLEM – 140

1. $\sqrt{27} + \sqrt{3} + \sqrt{12} =$

2. $\sqrt{3} = 1.7$ if $\sqrt{27} + \sqrt{3} = ?$

3. $\sqrt{2} = a, \sqrt{3} = b, \sqrt{5} = c, \sqrt{75} = ?$

4. $\sqrt{0.36} - \sqrt{0.25} =$

5. $\left(\sqrt{0.25} + \sqrt{0.36}\right) \div \left(\sqrt{1.44} + \sqrt{1.69}\right) =$

6. $\sqrt{9 + 2\sqrt{18}} =$

7. $\sqrt{11 + 2\sqrt{30}} =$

8. $\sqrt{10 + 2\sqrt{21}} =$

9. $\sqrt{12 - 2\sqrt{27}} =$

10. $\sqrt{8 + 2\sqrt{15}} - \sqrt{3} =$

11. $\left(\sqrt{3} - \sqrt{2}\right)^2 + \left(\sqrt{3} + \sqrt{2}\right)^2 =$

12. $\left(\sqrt{12} - \sqrt{7}\right)^2 + \left(\sqrt{12} + \sqrt{7}\right)^2 =$

13. $2+4+6+\ldots+24=$

14. $4+6+8+\ldots+36=$

15. $5!-4!$

16. $\dfrac{6!}{4!}$

17. $\dfrac{6! + 4!}{4! - 3!}$

18. $\dfrac{10! + 8!}{10! - 8!}$

19. $\dfrac{6! + 2!}{6! - 2!}$

20. $\dfrac{8! + 4!}{3! + 2!}$

PROBLEM – 141

1. $0.\overline{2} + 0.\overline{7} =$

2. $0.\overline{7} + 0.\overline{8} =$

3. $(\overline{0.4}) \times \frac{9}{4} =$

4. $0.1\overline{2} =$

5. $0.\overline{18} =$

6. $0.2\overline{4} =$

7. $4.\overline{2} + 5.\overline{6} =$

8. $0.\overline{11} + 0.\overline{22} =$

9. $4.0\overline{5} =$

10. $0.1\overline{3} + 0.2\overline{4} =$

11. Find the arithmetic mean of 4,8 and 12

12. Find the arithmetic mean of 2, 4,18 and 24

13. Find the arithmetic mean of 4,8,-6,12,28

14. Find the arithmetic mean of 4,6,8, 12

15. Find the arithmetic mean of 6,8,4,2,16

16. Find the median of -4,6,3,8,12

17. Find the mode of 2,3,1,7,3,6,3,8,3

18. Find the mode of 6,8,12,8,8,6,9,7

19. Find the mode of 2,9,7,9,7,8,9,7,9

20. Find the mode of 2,7,3,7,6,7,8,9,7,8,3

PROBLEM – 142

1. Find the range of: 3,27,86,100

2. Find the range of : 4,99,27,36,40

3. Find the range of: 2,-6, 36,42,30

4. Find the range of: 7,12,5,97

5. Find the range of: 6,-16,70,97,18

Find the next number in each sequence.

6. 32,30,28,26……

7. 48,44,40,36……

8. 3,9,27,81…..

9. 1,4,9,16,….

10. 1,8,27,64…..

11. 3,12,48……

12. 5,10,20,40,80?

Evaluating expressing.

13. Evaluate: x^2y+y^2 for x=4, y=2

14. Evaluate: $ab+ac+4$ for a=4, b=2, c=1

15. Evaluate: $a^2 b^2+ab$ for a=4, b=2

16. Evaluate: $(ab)^2+(bc)^3$ for a=1, b=2, c=3

17. Evaluate: $2x^3+3y^2+xy$ for x=1, y=2

18. Evaluate: $5x^3y^2$ if x=3, y=i

19. Evaluate: x^2-4x+7 if x=3

20. $4x^2yz^2 +6xyz+7$ if x=1, y=2, z=3

PROBLEM − 143

Equations.
*x+a=b, x=b-a

1. x+7=8

2. x+19=27

3. x+2.6=4.3

4. $x + 12\frac{1}{3} = 13\frac{1}{2}$

5. $x + \frac{1}{7} = \frac{1}{3}$

* x-a=b, x=a+b

6. x − 7=12

7. x − 9=27

8. x − 3.7=6.3

9. $x - \frac{1}{3} = \frac{1}{7}$

10. $x - 2\frac{1}{3} = 2\frac{1}{7}$

* $\frac{x}{a} = b$, $x = a \cdot b$

11. $\frac{x}{3} = 7$

12. $\frac{x}{12} = 7$

13. $\frac{x}{6 \cdot 2} = 4 \cdot 3$

14. $\frac{x}{7 \cdot 5} = 5 \cdot 2$

15. $\frac{x}{9} = 3 \cdot 3$

*$ax + b = cx + d$, $ax - cx = d - b$, $x = \dfrac{d-b}{a-c}$

16. 7x+2= 3x+22

17. 16x+16=6x+6

18. 9x-12=4x+18

19. 5x+5=2x+7

20. 12x-12=6x+6

PROBLEM – 144

Solving proportions for value.

* $\dfrac{x}{a} = \dfrac{b}{c}$, $x = \dfrac{ab}{c}$

1. $\dfrac{x}{3} = \dfrac{7}{2}$

2. $\dfrac{x}{5} = \dfrac{6}{9}$

3. $\dfrac{x}{6} = \dfrac{3.2}{4}$

4. $\dfrac{x4}{3} = \dfrac{2}{5}$

5. if 6x-6=6, then 8x-4

6. if 9x-9-=18 then 20x-7

7. if $x^2 = x+2$, then $x^3 = ?$

8. if $t^2 = t+6$, then $t^3 = ?$

9. 4x-4=12, then $4x^2-2$

10. $\dfrac{1}{2} + \dfrac{1}{4} + \dfrac{1}{8} =$

11. $\dfrac{1}{3} + \dfrac{1}{9} + \dfrac{1}{27} =$

12. $\dfrac{1}{6} + \dfrac{1}{12} + \dfrac{1}{36} =$

13. $\dfrac{1}{7} + \dfrac{1}{14} + \dfrac{1}{21} =$

14. $\dfrac{1}{10} + \dfrac{1}{20} + \dfrac{1}{30} =$

* $a^2 - b^2 = (a+b) \cdot (a-b)$

15. $9^2 - 3^2 =$

16. $44^2 - 36^2$

17. $63^2 - 42^2$

18. $199^2 - 188^2$

19. $(2003)^2 - (2002)^2$

20. $11^2 \times 7 - 7 \times 77 =$

PROBLEM – 145

1. $3^6 =$

2. $2^8 =$

3. $\left(\dfrac{2}{3}\right)^5 =$

4. $41^2 + 4^2 =$

5. $61^2 + 2^2 =$

6. $23^2 + 32^2 =$

7. The area of a right triangle with legs of length 15 and 20 is……

8. The perimeter of a right triangle with legs of length 10 and 24 is ……

9. The area of right triangle with legs and length (2x+2) and 6 is ….

10. The area of right triangle with legs and length (3x+4) and 8 is….

11. The sum of the terms in the arithmetic sequence 3,6,9,12…..+36 is……

12. The sum of the terms in the arithmetic sequence 5,10,15,20….+50 is….

13. How many triangle of figure?

14. How many triangle of figure?

15. How many rectangle of figure?

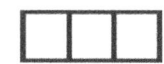

16. Find the all triangle perimeter.

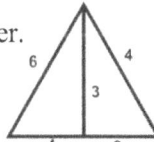

17. Find the all rectangle perimeter.

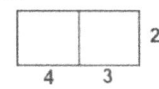

18. The sum of the terms in the arithmetic sequence 7,11,15,19,…..,51 is

19. $\dfrac{3}{8} \cdot 24 =$

20. $\dfrac{3}{8}$ of 80 is …..

PROBLEM − 146

1. 12_4 in base 10 is …..

2. 31_3 in base 10 is …..

3. 23_4 in base 10 is …..

4. 62_9 in base 10 is …..

5. 141_3 in base 10 is …..

6. 1203_4 in base 10 is …..

7. 121_3 in base 10 is …..

8. 21_4 in base 10 is …..

9. 21_9 in base 10 is …..

10. 1324_6 in base 10 is …..

11. 67_9 in base 10 is……

12. The number 232 written in base 9 is $(……..)_9$

13. The number of positive integral divisors of 80 is ………..

14. The number of positive integral divisors of 75 is ………..

15. The number of positive integral divisors of 90 is ………..

16. The number of positive integral divisors of 124 is ………..

17. The reciprocal of $\frac{12}{17}$ is …….

18. The reciprocal of $\frac{29}{28}$ is …….

19. The reciprocal of 13 is …….

20. The reciprocal of $2\frac{3}{5}$ is …….

PROBLEM – 147

1. If $3^a = m$, $3^{3a+3} = ?$

2. If $2^m = n$, $2^{2m+2} = ?$

3. If $2^a = m$, $3^a = n$, $72^a = ?$

4. The area of right triangle with hypotenuse of length 20 and one length of length 16 cm is…….

5. How much simple interest will an account earn in six years if $600 is invested at 4% interest per year…….

6. What is the simple interest on $6000 invested at an annual rate of 5% over four years…..

7. What is the simple interest on $3000 at 2% semiannual rate over five years…..

8. If six pounds of watermelon cost $3.25, at the same rate how much will 30 pounds of a watermelon last?

9. The sum of three consecutive integers is 39. What is the largest integer?

10. The sum of three odd consecutive integers is 45. What is the largest number?

11. The sum of three even consecutive integers is 54. What is the largest number?

12. The sum of four consecutive even integer is 92. What is the smallest integer?

13. The sum of six consecutive odd integer is 282. What is the smallest integer?

14. Two integer total 38. One integer is 14 larger than the other. What is the one of the two integers.

15. Two integer total 50. One integer is 20 larger than the other. What is the larger number

16. Two integer total 42, difference is 6/ Find the ratio

17. Two integer total 50, difference is 40. What is the smaller number?

18. What is the simple interest on a loan of $6000 for five years at an annual interest rate of 5%

19. Which digit is in the ten thousandths place 7462.89763

20. If the six books cost $24.48.What is the price each book.

PROBLEM – 148

1. Find of the GCF of 36 and 60.

2. The GCF of the expression 9m and 3m is...

3. The sum of two integer is -3. The product of two integer is -54. What are two integer

4. Two rational number have a sum of 22 and difference of 2. What are largest number?

5. The sum of two integers is 16. The product of two integers is 60. What are two integers?

6. Which number is between $\frac{1}{4}$ and $\frac{2}{5}$?

7. Which number is between $\frac{7}{3}$ and $\frac{3}{4}$?

8. What is the surface area of a cube with a side of length 4 inch?

9. What is the surface area of a cube with a side of length 10 cm

10. What is the surface area of a cube with a side of length 7 cm?

11. What is the surface area of a rectangular solid that has a length of 6 cm, width of 4 cm and height of 8 cm.

12. The volume of rectangular prism is 60 cm^3. The height is 3 cm and the width is 4 cm. What is the length?

13. In a survey of 300 student, 90 said liked the ice-cream. What percent is this?

14. A map uses a scale 1 inch: 21 miles. What distance is represented by $7\frac{1}{2}$ inches on the map.

15. If the train speed is 42 mile per hour and the train travels for 5 hours. How far will the train travel?

16. Two angles in a triangle measures 42° and 64°. What is the measure of the third angle?

17. Two angles in a triangle measures 75° and 40°. What is the measure of the third angle?

18. Jack bought 7 math books for $ 49.28. What is cost each book?

19. Jack test scores are 92, 96, 91 and 98. Find test score average?

20. What is the sum of 4 hours 45 minute, 3 hours and 35 minute.

PROBLEM – 149

1. After school math club are 24 students. 11 of the student are boys. What is the ratio of girls to the total number of student?

2. Ahmet has 8 Algebra and 6 geometry book. What is ratio of the number of geometry book to the total number of the book.

3. Book store sold 36 Algebra and 27 geometry book. What is the ratio of total number of geometry sold to the number of the Algebra.

4. At the book store, there are 42 Turkish book and 27 Germany book. What is the ratio of Turkish book to Germany book.

5. Jack spent $45 for a book. How much does sack pay for 4 book?

6. Two dozen pencil cost $12.How much does cost 4 pencil?

7. Ahmet walked 24 mile in 6 hours. At that rate, how many miles can Ahmet walk in 4 hours.

8. Ahmet earned $96.00 and 6 hours. How much Ahmet earn in $18\frac{1}{2}$ hour.

9. Ahmet earned $84 and 7 hours. How much Ahmet earn in $6\frac{1}{2}$ hour.

10. Math book that original cost $84. In on sale for 10% off. How much will a costume save.

11. What decimal in simpliest form are equivalent to 44%.

12. What fraction is equivalent to 0.68.

13. Jack answered $\frac{7}{8}$ of the question on his test correctly. What percent he did not answer?

14. The rectangular garden has a length of 3x+4 and width of 2x+2. What expression represents the perimeter of the garden.

15. A square pool has a length of 2x+3. Find the perimeter.

16. Jack bought $2\frac{2}{3}$ pounds of watermelon for $ 1.30 per pound. How much did he spend in all.

17. A book store sale 300 each day. Six percent of the these books are geometry. How many geometry book does sale each day?

18. What is 24% of 10^3 ?

19. Jack wants to buy a computer that cost $350. He has saved 7% of the total cost of the computer. How much money has Jack saved?

20. Is it possible for the side length 2 cm, 3 cm and 8 cm to form a triangle…

PROBLEM − 150

1. Is it possible for the side length 7 cm, 9 cm and 10 cm to form a triangle.

2. Two sides of the triangle have measures of 8 cm and 10 cm. What is the perimeter of the triangle?

3. A triangle one angle is measures 64°? another angle measures 32°. What is the measure of the third angle of the triangle.

4. Find the area of a square that has sides 3 cm long.

5. Find the perimeter of a square that has sides (2x+3) long.

6. A rectangular prism has a volume of 420 cm^3/ Its height is 7 cm. What is its base are?

7. Two sides of a triangle have measures of 8 cm and 10 cm. What could be the length of the third side?

8. ABC is triangle. $\angle A = 60°$, $\angle B = 70°$ Compare triangle side?

9. ABC is triangle. AB=6cm, BC=8cm, AC=12cm. Compare triangle angles.

10. 12% of the eleventh grade at high school ride own car. What fraction of the eleventh grade do not used own car.

11. Jack has $32.60. He spent $\frac{2}{5}$ of it for books. How much did the books cost?

12. If 12 math book are sold for $15.60 each. What are the total sales for the books.

13. Jack spends $48.24 on 4 math book that each cost the same amount. What is the price of each book.

14. What is $\frac{9}{20}$ written as a percent?

15. What is 96% written as a fraction in simplest form?

16. Jack has owns 80 books 20% are math books. How many math books does he own.

17. Bank give Jack a loan for $2000 at a rate 6% interest for one year. How much interest will he pay for the loan.

18. Ahmet has $640 to deposit at a rate of 4%. What is the interest earned after one year?

19. Jack has $420 at an interest rate of 3%. What is the interest earned after other one year.

20. Ahmet writes the expressions $2^{-4} \times 2^6$. What is the product?

PROBLEM – 151

1. 8 is 4% of what number?

2. 24 is 6% of what number?

3. The length of a rectangle is 6 cm more than the width. The perimeter is 32 cm find the are?

4. The length of a rectangle is 11 cm less than width. The perimeter is 50 cm. Find the area?

5. Write 0.000021 is scientific notation.

6. The sum of the squares of two consecutive even integers is 100. Find the two integers.

7. The sum of the squares of two consecutive even integers is 244. Find there number averages.

8. Add $\frac{2}{3}$ to the product of 3 and $\frac{1}{5}$

9. Add $-\frac{2}{5}$ to the product of -10 and $\frac{7}{25}$

10. Add $-\frac{2}{9}$ to the product of -7 and $\frac{5}{21}$.

11. The sum of three consecutive odd integer is 129.What is the largest of them.

12. What is 80 divisible by?

13. Twice a number less 3 is equal to four times the sum 5 and that number. What is the number?

14. The sum of three consecutive even integer is 72. What are they?

15. The length of a rectangle is 4 more than twice its width. The perimeter is 96 cm. what are the area of a rectangle?

16. The similar triangles have side in the ratio $\frac{7}{9}$. What is the ratio of the areas of the triangles?

17. What is the base of a triangle of area 24 and height 8 cm.

18. The angle of a right triangle are in the ratio 2:4:6. If the smallest side of the triangle is 60 cm. What is the perimeter of triangle?

19. What is the number of square unit in the perimeter of an isosceles right triangle with area of 50 cm^2.

20. If a 45:45:90 triangle has a area 200 cm^2. What is the perimeter of triangle.

PROBLEM – 152

1. $5\frac{1}{2}$ gallon paint covers 300 square feet. How many square feet will $9\frac{3}{5}$ gallons of point cover.

2. The measures of the three angles of a triangle are $(2x)°$, $(4x)°$ and $(6x)°$ .What is the great angle?

3. Write the number 3.4×10^4 in standard form.

4. Write $3\frac{3}{8}$ as a decimal form.

5. What is the scientific notation for a length of 0.00000000124

6. A square garden has an perimeter of 100 cm. Find the area of garden.

7. A computer repairmen charges $45 plus $30 per hour. Write equation for this situation.

8. A rectangle is 24 cm long. The perimeter is 84 cm. find the area of rectangle.

9. Two legs of a right triangle each measure $(2x+4)$ cm. Find the area of triangle.

10. Jack has $3000 m a saving account that pay 3% simple interest. How much interest is earned on this money first year?

11. A cube has a total surface are of 600 square cm. What is the volume of cube?

12. Jack ate dinner at a restaurant. The total bill was $84. The tip was 12% What was the amount at the tip?

13. The ratio of girls to boys in a school is $\frac{7}{9}$. There are 320 students at school. How many boys are this school?

14. Math book original price is $60. Jack bought this book $45. By what percentage is the price of book reduced.

15. The price of a geometry book that regularly sells for $44 is reduced to $33. By what percentage is the price reduced.

16. Jack have 288 books. $\frac{3}{8}$ of the books are math books. How many of the books are math?

17. Jack bought new math books. He paid $40.80. The sale tax is 7%. What is the total cost including tax.

18. Ahmet bought new geometry books. He paid $70.80. The sale tax is 9%. What is the sale tax amount.

19. The area of circle is 8π square meters. Find the diameter of circle.

20. The area of circle is 12π square meter. Find the radius of circle.

PROBLEM – 153

1. Algebraic expressions.
   $24xyz+15x^2y+18yx$ What factor is common to all three term.

2. $6x^2y+7xy+xym$ Find the numerical coefficients of each term.

3. Find the perimeter of a rectangle with sides measuring 4.3 and 12.7 cm.

4. Seven less than 4 times a number is 21. Find the number.

5. Nine less than 5 times a number is 26. Find the number.

6. Jack has a test store of 94, 82 and 81 points. What must he score on the fourth exam to have an average score of at least 88 points.

7. A $40 math book cost $44.20 with sale tax. Find the tax rate.

8. The rectangle is 4 times as long as it is wide and its perimeter is 100 cm. Find its are.

9. One number is 4 times another and their sum is 35. Find the product numbers.

10. If a truck can travel 81 mile on 3 gallon of gas, how much gas will it need to travel 324 miles?

11. A and B being positive integers if A:B= $\frac{9}{11}$, what is the minimum value of A+B.

12. The ratio of two number is $\frac{7}{9}$. If the smaller number is 28. Find the greater number.

13. How much does a customer who buys 300 gram of honey at $15.75 per by need to pay?

14. 20 less than $\frac{2}{5}$ of a number is 20. What is the number?

15. The sum of $\frac{2}{3}$ of a number and $\frac{1}{5}$ of the same number is 30. Find the number.

16. Ahmet`s age is m and Jack`s is n years old. What will the sum of their ages be in 3t years?

17. If the arithmetic mean of 18 number is equal to 32, what is the sum of these numbers.

18. What is the percentage increase of on item sold for $240 Which is originally priced at $300?

19. What is the percentage increase from 160 to 200?

20. Find the sum of all composite numbers between 17 and 29.

PROBLEM – 154

1. Six algebra and three geometry books cost $75, while two algebra and five geometry books cost $125. Find the one algebra and geometry book cost?

2. If the sum of two numbers is 32 and the difference is 2. Find the larger number square?

3. If the sum of two numbers is 20 and the difference is 4. Find the product numbers.

4. If the product of two consecutive ever number is 120. Find the smaller number square.

5. The square of a number plus 6 is to 55. Find the number cube.

6. A square of a number plus half its number is equal to 105. Find the number cube.

7. The cube of a number plus 6 is equal to 70. Find the number square.

8. If the sum of two even numbers is 22 and the product of two numbers is 120. Find the number ratio.

9. One number is 5 times another number if their sum is 30. Find the difference number.

10. One number is 5 times another number if their sum is 26. Find the difference number.

11. Find the sales tax on a computer cost $300 if the rate is 4%.

12. New math book originally price is $60? And was discount 22% for a sale, what was the discount price

13. If 6 geometry books cost $18.30. How much will a geometry books cost.

14. Which of the following equations represent a proportional function.
    $y=8$ $y=8x$
    $y=6+x$ $y=x^3$

15. Jack has $60 in a saving account. He will start adding $8 to his account each day. Write equation, in his account after x days.

16. Which side lengths turn a right triangle.
    I ; 2,4,6   II ; 6,8,10
    III ; 3,5,7   IV; 1 , 3 5

17. What is the side length of a cube that has a total surface area of 54 square centimeter.

18. A right triangle has legs that measure 2 cm and 4 cm. What is the length of the hypotenuse in centimeter.

19. A right triangle has legs that measure 5 cm and 12 cm what is the perimeter of triangle.

20. The measures of the three angles of a triangle are given by 2x+2, 4x+3, x+5. What is the measures of the smallest angle?

PROBLEM – 155

1. Find the total cost of 4 pairs of algebra books at $48.72 each and 7 pairs of geometry books at $21.35 each.

2. Jack bought 4 pairs of chemistry books at $24.72 each. If he paid for them with a $100 bill, how much change did he receive?

3. A rental car driver charges $12 plus $3.72 per mile find the total cost of a 9 mile trip.

4. Jack earns $72000 a year, what is the Jack`s 3 month salary.

5. School principle said ¾ of student enrolled for next school year. If the 800 student study school. How many student did not enrollment?

6. What percent of 35 is 25

7. 24 is what percent of 40

8. New school has 900 student if 60% of them are girls. How many of them student are boys?

9. New math book cost $28 and was discounted $7. What was the discount rate?

10. If tile sells for $12.18 per square. How much will it cost to cover a kitchen that measure 24.12 square.

11. Jack earns $7224 in 4 month, how much will he earn in $2\frac{1}{2}$ years

12. Find the Celsius temperature when the Fahrenheit temperature is 96. (use C= $\frac{5}{9}$ (F – 32)

13. University has 775 student if there are 75 more boys than girls. Find the ratio boys to girls.

14. Find the interest on a loan of $6000 at 3 % for 7 years.

15. If Jack saves $82 for 2 month. How much will Jack have saved in $3\frac{1}{2}$ year.

16. Find the range of : 6,18,24,3,42

17. Find the mode of: 3,14,12,12,18,19

18. A right triangle has legs that measure 15 cm and 20 cm. What is the length of the hypotenuse in centimeter.

19. Find the median of : 6,3,5,8,12,16,24

20. A right triangle has legs that measure 9 cm and 12 cm. What is perimeter of the triangle.?

PROBLEM − 156

1. A right triangle has legs that measure 18 cm and 24 cm. What is perimeter of triangle.

2. ∠A and ∠B are vertical angles. If the measure of ∠B is 67, find the measure of ∠A.

3. ∠M and ∠N are supplementary angles. If the measure of ∠M is 68°. Find the measure of ∠N.

4. ∠A and ∠B are complementary angles. If the measure of ∠A is 28°. Find the measure of ∠B.

5. ∠M and ∠N are complementary angles. If the ∠M is 10x-5 and ∠N= 6x+20. Find ∠M

6. ∠M and ∠N are supplementary angles. If the ∠M is 3x+2 and ∠N=5x-8. Find ∠M.

7. If the measure of an angle is 15°, find the measure of its supplement.

8. If the measure of an angle is 42°, find the measure of its complement

9. ∠A and ∠B are vertical angles. If ∠A= 4x+14, ∠B=5x−12. Find ∠A.

10. Find the midpoint of $\overline{KL}$ if K(4,8) and B(-2,-12).

11. If M is midpoint of $\overline{AB}$, find the coordinates of B if A(12,-4) and M(8,6).

12. The cost of tile is $0.36 per square cm. How much will it cost to the square 25 cm by 25 cm.

13. Jack went paint front garden wall. Front garden wall that is 15 m long and 12 meter wide. What is the area of the painting?

14. A triangle has a base length of 12 cm and a height of 9 cm. What is the ratio area to perimeter.

15. Two side of a triangle measure 12 cm and 8 cm. Write inequality show to possible lengths for the third side x.

16. ABC is triangle. AB=6, AC=8 and BC=10 cm Write relationship the angles in ABC.

17. ABC is triangle. ∠A=60, ∠B=70. Write relationship to angle sides in ABC.

18. Find the area of an equilateral triangle with side length 12cm?

19. Find the number of sides of a polygon whose sum of the measure of its interior angles is 2160.

20. ABCD is a square AB=4x-2 and BC=2x+10 cm. Find the square area?

PROBLEM – 157

1. The ratio of two complementary angles is 1:2. Find greatest measure of greatest angle.

2. The ratio of two supplementary angles is 1:5. Find the difference angle.

3. The ratio of the measure of the angle in a triangle is 2:4:6. Find the smallest angle.

4. The ratio of the measures of the sides of rectangle is 7:9. Find the minimum perimeter of rectangle.

5. ABC is triangle. The ratio of sides 1:3:5. If the perimeter of the triangle is 27.90 cm, find the small side.

6. The ratio of the measures of the three angles in triangle is 1:2:3. Big angle is 90° Find the small angle.

7. ABC is triangle AB=AC, $\angle A=70°$, find the $\angle C$.

8. ABC is triangle AB=AC, $\angle B=72°$, find the $\angle A$.

9. The ratio of the measures of the sides of a triangle is 2:3:5. Big side is 15 cm. Find the perimeter of the triangle.

10. The ratio of the measures of the 3 sides of a triangle is 1:3:5. Small side is 2x+2. Find the perimeter of the triangle.

11. The ratio of the measures of the parallelogram angle is 2:3. Find the small angle.

12. The ratio of the sides of a parallelogram angle is 2:3. If the perimeter of the parallelogram is 60 cm, what is the length of the smallest side?

13. The ratio of the angles in parallelogram angle is 1:3. What is the measure of the smallest angle.

14. ABC is triangle AB:AC=BC $\angle A$ =5x+5, find x value.

15. ABC is triangle. $\angle A=\angle B=\angle C$, AB=2x+3, find the perimeter of triangle.

16. Write an equation parallel to x-2y=4, that passes through the point (5,3)

17. Write an equation perpendicular to $y = \dfrac{x}{3} + 3$ the passes through to point (2,7)

18. Write a linear equation that passes through the point (-4,3) with a slope of 1/3.

19. Write a linear equation that passes through the point (5,2) with a slope of 2/3.

20. Write a linear equation that passes through the point (2,5) and (-3,7).

PROBLEM – 158

1. Find the surface area of a rectangular prism with side of length 2,3 and 5 cm

2. Find the surface area of a rectangular prism if the base is square with edge length 7 cm, and the height is 5 cm.

3. Find the surface area of a cube with length 12 cm.

4. Find the surface area of a cube with length 2.2 cm

5. Find the volume of a cube with length 3.2 cm.

6. In $\triangle ABC$, $\angle A=105°$, $\angle B=45°$ and AC=12cm. Find the area at the ABC

7. In ABC, AB=AC=20cm, and BC=24cm. Find the area of the ABC.

8. Find the area of an equilateral triangle with side length 8 cm.

9. Find the number of sides of a polygon, if the sum of the measures of its interior angles is 2160°

10. The ratio of the interior angles of a quadrilateral is 1:2:3:4. Find the measure of the largest angle.

TEST ANSWERS

Test No	Q1	Q2	Q3	Q4	Q5	Q6	Q7	Q8	Q9	Q10
Test-1	D	A	B	B	A	D	A	A	B	A
Test-2	D	B	A	D	A	C	B	B	A	D
Test-3	B	C	E	E	E	O	B	E	E	E
Test-4	B	A	A	B	D	D	C	A	D	C
Test-5	A	A	A	D	D	D	B	A	C	B
Test-6	B	B	D	A	C	D	A	B	B	C
Test-7	B	B	C	D	C	B	A	B	D	D
Test-8	C	B	A	D	D	D	A	A	D	A
Test-9	D	B	A	A	C	C	C	C	B	C
Test-10	A	D	C	B	C	C	A	D	A	B
Test-11	B	B	D	C	C	D	A	B	C	D
Test-12	B	D	C	D	C	A	B	C	B	A
Test-13	A	C	B	C	C	B	C	C	A	C
Test-14	B	B	A	C	D	D	C	D	A	C
Test-15	B	A	D	C	C	C	A	C	A	A
Test-16	A	B	C	D	C	A	C	A	B	B
Test-17	C	D	B	A	D	C	C	D	A	C
Test-18	D	C	B	B	B	C	C	D	A	A
Test-19	B	A	D	C	D	A	A	B	D	A
Test-20	A	B	B	D	B	C	A	B	C	D
Test-21	B	A	B	A	D	A	A	C	D	D
Test-22	B	B	C	D	D	B	D	A	D	C
Test-23	A	B	D	D	C	B	C	D	D	B
Test-24	C	B	D	A	C	C	D	B	A	C
Test-25	D	D	C	A	A	D	D	D	B	D
Test-26	C	C	D	C	A	D	D	A	D	C
Test-27	B	C	D	D	A	B	A	C	B	A
Test-28	D	C	D	A	D	C	D	D	C	D
Test-29	A	B	C	D	B	A	D	A	A	D
Test-30	C	A	C	A	B	A	A	A	D	D
Test-31	B	A	D	C	D	C	A	D	C	B
Test-32	B	D	A	D	D	C	C	A	A	B
Test-33	B	A	C	A	C	B	B	A	A	D
Test-34	A	B	B	A	B	B	A	D	D	C
Test-35	C	C	B	A	B	C	B	D	C	D
Test-36	B	C	A	D	D	A	A	C	D	A

Test-37	B	B	C	C	C	B	D	A	D	A
Test-38	C	B	D	A	B	C	D	C	C	B
Test-39	A	C	D	C	C	D	B	D	D	C
Test-40	A	C	D	C	D	C	B	B	D	B
Test-41	C	C	D	A	D	D	D	D	B	A
Test-42	C	D	C	B	C	A	D	D	A	C
Test-43	A	A	C	C	A	C	A	D	B	D
Test-44	B	D	B	D	A	B	D	D	B	C
Test-45	D	C	A	D	C	A	C	C	A	A
Test-46	A	D	A	C	C	D	A	A	B	D
Test-47	D	C	B	A	A	B	C	D	D	A
Test-48	A	C	A	C	C	B	A	D	B	A
Test-49	C	C	C	B	D	A	C	D	D	D
Test-50	B	D	D	A	D	C	A	C	C	D
Test-51	B	B	B	A	D	A	C	C	C	D
Test-52	B	A	C	A	D	D	B	D	D	A
Test-53	B	B	A	D	C	C	D	C	B	D
Test-54	B	A	B	C	D	A	A	A	B	D
Test-55	A	B	C	D	A	D	D	D	D	D
Test-56	A	C	D	A	D	A	C	A	D	A
Test-57	B	B	B	D	A	B	D	C	A	C
Test-58	B	D	A	C	A	B	A	D	D	B
Test-59	A	D	B	B	B	D	B	D	A	A
Test-60	B	A	D	B	C	A	A	D	C	D
Test-61	D	B	A	C	B	B	C	D	C	A
Test-62	D	A	B	A	D	C	D	D	D	C
Test-63	D	A	C	D	B	A	B	D	A	C
Test-64	C	A	D	A	D	A	C	D	D	C
Test-65	C	A	D	A	D	A	C	D	D	C
Test-66	A	D	A	A	D	C	C	A	D	D
Test-67	A	A	B	A	A	B	A	D	D	A
Test-68	A	B	D	D	C	B	A	A	D	D
Test-69	D	B	B	B	B	D	D	B	A	D
Test-70	D	A	C	D	B	D	D	A	C	B
Test-71	D	B	D	C	D	B	D	A	D	D
Test-72	B	D	A	D	C	C	A	A	A	B
Test-73	A	D	C	C	A	D	B	D	B	B
Test-74	D	C	D	C	C	A	B	D	C	A

Test-75	B	A	D	A	D	C	C	B	C	C
Test-76	A	B	A	A	C	B	B	B	A	C
Test-77	A	D	D	D	D	C	C	C	A	D
Test-78	D	D	C	A	A	D	B	C	A	A
Test-79	A	A	B	D	B	D	D	A	A	C
Test-80	B	C	B	C	D	A	C	B	A	A
Test-81	D	D	A	B	A	C	D	B	B	D
Test-82	B	C	A	D	C	D	A	C	A	C
Test-83	B	A	C	A	C	D	D	C	A	D
Test-84	B	B	D	A	C	A	C	B	B	C
Test-85	C	D	D	D	A	D	C	B	A	D
Test-86	C	A	B	D	B	B	C	B	D	D
Test-87	D	D	D	A	C	C	D	D	D	A
Test-88	C	A	A	C	A	A	D	A	D	B
Test-89	B	B	C	D	C	A	A	A	C	C
Test-90	B	A	B	A	D	D	C	D	D	C
Test-91	B	B	C	B	C	D	D	D	C	D
Test-92	C	D	A	A	A	B	A	A	C	D
Test-93	C	B	B	A	D	D	C	B	C	A
Test-94	A	D	C	B	A	C	C	A	D	D
Test-95	B	D	A	A	B	C	B	D	C	C
Test-96	D	B	A	D	C	C	C	C	A	C
Test-97	A	D	D	A	A	D	A	C	D	A
Test-98	A	C	D	C	A	B	C	D	C	A
Test-99	D	A	C	C	D	D	C	C	A	D
Test-100	C	B	A	B	D	A	C	C	B	D
Test-101	B	A	C	D	B	D	A	C	B	A
Test-102	C	B	A	D	A	C	B	C	D	A
Test-103	B	B	D	A	D	D	D	B	A	B
Test-104	D	C	A	B	B	A	C	A	B	D
Test-105	A	B	A	B	B	C	A	D	D	A
Test-106	B	C	C	B	A	D	D	D	A	B
Test-107	C	C	B	D	C	C	C	A	A	D
Test-108	D	A	B	A	D	D	B	B	A	C
Test-109	D	C	A	D	B	B	B	B	B	C
Test-110	A	D	D	B	A	C	A	B	C	D
Test-111	D	D	B	A	B	B	B	A	A	D
Test-112	A	C	B	A	D	B	C	C	D	A
Test-113	A	B	A	B	C	A	A	D	A	B
Test-114	B	B	D	A	B	B	B	A	C	A
Test-115	D	C	C	B	C	D	A	B	D	C

PROBLEMS' ANSWERS

	P-116	P-117	P-118	P-119	P-120	P-121	P-122	P-123	P-124	P-125
Q-1	2997	1088	22	0	156	270	32	8	330	289
Q-2	94	1088	33	1	1632	240	54	4	950	361
Q-3	258	3964	12	1	171	39	145	6	320	441
Q-4	222	811	6	2	72	3700	33	7	650	25/49
Q-5	552	1116	259	1	256	29	496	9	770	1024
Q-6	2660	1243	162	2	1	280	68	7	9800	16/81
Q-7	176	2664	757	1	144	123	113	9	12400	343/125
Q-8	1099	972	249	0	44	321	163	3	9800	421
Q-9	2670	1422	24	0	22	27	683	7	5600	313
Q-10	289	5593	206	2	800	674	2700	8	1200	194
Q-11	848	652	4	1	324	160	41	193	0.49	30
Q-12	5340	5664	32	0	108	6250	229	102	0.12	36
Q-13	1082	864	14.2	0	5184	12500	24	254	0.99	17
Q-14	45	1221	9	0	9/4	25	984	96	0.8	19
Q-15	191	1332	30	2	75	100	113	130	0.4	140
Q-16	150	801	32	8	16	3600	172	176	3280	280
Q-17	3	801	163	4	1	64000	11	264	6500	41
Q-18	96	2172	146.5	2	1	8000	120	214	726	605
Q-19	678	5338	332	23	252	6250000	10700	158	484	68
Q-20	3996	39403	121	4	1	216000	5100	54	972	219

	P-126	P-127	P-128	P-129	P-130	P-131	P-132	P-133	P-134	P-135
Q-1	402	32	6	>	20.18	19/30	2880	24	1/3	4.8
Q-2	-18	125	5	>	37	0.75	144	42	3/5	55/54
Q-3	214	216	7	<	171	7/9	336	60	7/5	1.4
Q-4	111	27	5	>	108	3/10	180	36	¾	48
Q-5	857	36	16	>	103	1	720	204	7/4	1/72
Q-6	1320	288	12	>	132	11/45	14	48	6/7	28/3
Q-7	432	625	10	<	167	0.325	35	60	¾	4
Q-8	1072	200	21	<	165	3/10	1460	175	4/3	8/3
Q-9	1936	128	12	>	454	5/4	15	90	2/3	14/3
Q-10	980	81	84	<	581	15/28	2	144	8/9	2
Q-11	557	192	2.33	1024	5	31/12	159	240	6.16	2/3
Q-12	1050	108	1.3	1936	10	11/8	448	2.6	7.81	2
Q-13	1140	98	5.4	3025	25	31/21	336	0.36	5.36	14
Q-14	1785	96	5.3	5929	8	1.7	8	1.23	7.81	7/18
Q-15	960	65.3	10.7	562	30	51/14	11	3.6	9.2	18/11
Q-16	1316	39	3.21	460	11.3	1.7	42	3.24	5	45/7
Q-17	1092	64	5.08	434	1	19/24	72	0.024	7.58	52/55
Q-18	513	28.5	12.08	30	10.25	311/154	7	0.00144	17.58	14/3
Q-19	1408	44	15.6	10	100	17/21	5	0.064	6.59	6/5
Q-20	324	972	10.8	100	4	19/21	6	9	6.31	3/7

	P-136	P-137	P-138	P-139	P-140	P-141	P-142	P-143	P-144	P-145
Q-1	7/13	0.25	12	8×10^6	$6\sqrt{3}$	1	97	1	10.5	729
Q-2	13/9	0.75	12	15×10^7	6.8	17/9	95	8	3.33	256
Q-3	39/28	0.35	12	21×10^{13}	5b	1	48	1.7	4.8	32/729
Q-4	1	0.68	16.2	20.8×10^6	0.1	4/33	92	1.5	0.3	1697
Q-5	98/39	0.76	36	21×10^9	11/25	2/11	113	4/21	12	3725
Q-6	45/22	72%	33%	3000	$\sqrt{3}+\sqrt{6}$	8/33	24	19	53	1553
Q-7	13/29	64%	33%	2000	$\sqrt{6}+\sqrt{5}$	8.9	32	36	3x+2	150
Q-8	35/33	9%	20%	30000	$\sqrt{7}+\sqrt{3}$	0.33	243	10	7t+6	60
Q-9	4/3	60%	8.30%	1060000	$3-\sqrt{3}$	4/33	25	10/21	62	6X+6
Q-10	5/7	84%	11%	60	$\sqrt{5}$	37/90	125	94/21	7/8	12X+6
Q-11	7/10	24%	33%	29	10	88	192	21	13/27	234
Q-12	9/10	36%	20%	37	38	12	160	84	5/18	275
Q-13	12/100	80%	20%	38	156	9.2	36	26.66	11/42	3
Q-14	64/100	96%	33%	44	340	7.5	16	39	11/6	6
Q-15	123/1000	12%	200%	-16	79	7.2	72	29.7	72	6
Q-16	71/100	37.50%	300%	36	30	6	220	5	640	40
Q-17	64/100	75%	400	35	124/3	3	16	-1	2205	40
Q-18	39/50	45%	25%	41	91/89	8	-135	6	2057	342
Q-19	98/100	19%	33%	51	361/359	9	4	0.66	4005	9
Q-20	424/1000	233%	11%	61	5043	7	115	3	308	30

	P-146	P-147	P-148	P-149	P-150	P-151	P-152
Q-1	7	$27m^3$	12	13/24	Yes	200	524
Q-2	10	$4n^2$	3m	3/7	35	400	90
Q-3	11	n^2m^3	(-9.6)	0.75	84	55	34000
Q-4	56	96	12	14/9	9	126	3.375
Q-5	22	24	6--10	180	4x+12	21/1000000	12.4×10^{-10}
Q-6	99	1200	<	0.5	60	6--8	625
Q-7	16	300	37/12	16	17	11	y=30x+45
Q-8	9	16.25	96	111	ab>bc>ac	19/15	432
Q-9	19	18	600	78	B>A>C	-3.2	$2x^2+8x+8$
Q-10	340	17	294	8.4	88%	-1.8	90
Q-11	6	20	208	0.44	13.04	45	1000
Q-12	277	20	5	17/25	187.2	10	10,08
Q-13	10	41	30%	12.50%	12.06	-11.5	180
Q-14	6	12--26	73.5	14x+12	45%	22,24,26	25%
Q-15	12	35	210	8x+12	24/25	572	25%
Q-16	4	24	74	3.46	16	49/81	108
Q-17	17/12	5	65	18	120	6	43.66
Q-18	28/29	1500	7.04	240	25.6	$36\sqrt{3}$	6.37
Q-19	1/13	7	94.25	24.5	12.6	$20+10\sqrt{3}$	8
Q-20	5/13	4.8	(8.20)	No	2^{-8}	$40+20\sqrt{3}$	$2\sqrt{3}$

	P-153	P-154	P-155	P-156	P-157	P-158
Q-1	3xy	25	15.23	72	60	62
Q-2	6, 7, 1	17	1.12	67	120	238
Q-3	34	96	45.48	112	30	864
Q-4	7	100	18000	62	32	29.04
Q-5	7	343	200	42	3.1	32.768
Q-6	95	1000	140%	71.75	30	$18+18\sqrt{3}$
Q-7	0.11	16	60%	165	55	192 cm^2
Q-8	400	5/6	360	48	36	$16\sqrt{3}$ cm^2
Q-9	196	20	25%	118	30	9
Q-10	12	4	294-	(1.-2)	18X+18	144
Q-11	20	12	54180-	(4.16)	72	
Q-12	36	46.8	35-	225	12	
Q-13	4.725	3.05	14/17	180	45	
Q-14	200	Y=8x	1260	1.5	11	
Q-15	450/13	y=8x+60	1724	4<x<20	6X+9	
Q-16	m+n+6t	II	39	A>B>C	y=x/2 +1/2	
Q-17	576	3	12	AC>BC>AB	y=-2x+11	
Q-18	25%	$2\sqrt{5}$	$2\sqrt{5}$	S	y=x/3 +13/3	
Q-19	20%	30	8	17	y=2x/3 -4/3	
Q-20	227	50.6	36	484	y=-2x/5 +7	

About the Authors

Veysel Dereli

Veysel Dereli graduated from Hacettepe University in Turkey in 1996 with earning a Bachelor's Physic Teacher. Mr. Dereli has 10+ years of physics teaching experience with advanced high school students. He is a physics and test prep expert who has been teaching in learning centers and high school test since 1996. He taught high school physics. Also he has been coaching physics team in the school for four years. He lives in Houston.

Tayyip Oral

Tayyip Oral is a mathematician and test prep expert who has been teaching in learning centers and high school test since 1998. Mr. Oral is the founder of 555 math book series which includes variety of mathematics books. Tayyip Oral graduated from Qafqaz university with a Bachelor`s degree in Industrial Engineering. He later received his Master`s degree in Business Administration from the same university. He is an educator who has written several SAT Math, ACT Math, Geometry, Math counts and Math IQ books. He lives in Houston,TX.

Books by Tayyip Oral

1. Sheryl Knight, Mesut Kizil, Tayyip Oral, ACCUPLACER MATH PREP, (1092 Questions with Answers), 2018

2. Sheryl Knight, Mesut Kizil, Tayyip Oral, TSI MATH Texas Success Initiative, (1092 Questions with Answers), 2017

3. Tayyip Oral, Osman Kucuk, Hasan Tursu, Geometry for SAT & ACT (555 Questions with Answers), 2017

4. Tayyip Oral, Ferhad Kirac, Bekir Inalhan, Algebra for The New SAT, Level – 1, (1111 Questions with Answers), 2017

5. Sheryl Knight, Tayyip Oral, Servet Oksuz, Algebra for ACT, Level – 1, (1080 Questions with Answers), 2017

6. Kristin Alexander, Tayyip Oral, Sait Yanmis, 555 Gifted and Talented, Question Sets for the Mathematically Gifted Middle Grade Scholar (1111 Questions with Answers), 2017

7. Steve Warner, Tayyip Oral, Sait Yanmis, 1000 Logic&Reasoning Questions for Gifted and Talented Elementary School Students, 2017

8. Tayyip Oral, 555 ACT Math, 1110 Questions with Solutions, 2017

9. Tayyip Oral, 555 Math IQ for Elementary School Students (1270 Questions with Answers), Second Edition, 2017

10. Tayyip Oral, Ersin Demirci, 555 SAT Math, 2016

11. Tayyip Oral, Sevket Oral, 555 ACT Math - II, 555 Questions with Answers, 2016

12. Tayyip Oral, 555 Geometry (555 Questions with Solutions), 2016

13. Tayyip Oral, Dr. Steve Warner. 555 Math IQ Questions for Middle School Students: Improve Your Critical Thinking with 555 Questions and Answer, 2015

14. Tayyip Oral, Dr. Steve Warner, Serife Oral, Algebra Handbook for Gifted Middle School Students, 2015

15. Tayyip Oral, Geometry Formula Handbook, 2015

16. Tayyip Oral, Dr. Steve Warner, Serife Oral, 555 Geometry Problems for High School Students: 135 Questions with Solutions, 2015

17. Tayyip Oral, Sevket Oral, 555 Math IQ questions for Elementary School Student, 2015

18. Tayyip Oral, 555 ACT Math, 555 Questions with Solutions, 2015

19. Tayyip Oral, Dr. Steve Warner, 555 Advanced math problems, 2015

20. Tayyip Oral, IQ Intelligence Questions for Middle and High School Students, 2014

21. T. Oral, E. Seyidzade, Araz publishing, Master's Degree Program Preparation (IQ), Cag Ogretim, Araz Courses, Baku, Azerbaijan, 2010.
    A master's degree program preparation text book for undergraduate students in Azerbaijan.

22. T. Oral, M. Aranli, F. Sadigov and N. Resullu, Resullu Publishing, Baku, Azerbaijan - 2012 (3.edition)

    A text book for job placement exam in Azerbaijan for undergraduate and post undergraduate students in Azerbaijan.

23. T. Oral and I. Hesenov, Algebra (Text book), Nurlar Printing and Publishing, Baku, Azerbaijan, 2001.

    A text book covering algebra concepts and questions with detailed explanations at high school level in Azerbaijan.

24. T.Oral, I.Hesenov, S.Maharramov, and J.Mikaylov, Geometry (Text book), Nurlar Printing and Publishing, Baku, Azerbaijan, 2002.

    A text book for high school students to prepare them for undergraduate education in Azerbaijan.

25. T. Oral, I. Hesenov, and S. Maharramov, Geometry Formulas (Text Book), Araz courses, Baku, Azerbaijan, 2003.

    A text book for high school students' university exam preparation in Azerbaijan.

26. T. Oral, I. Hesenov, and S. Maharramov, Algebra Formulas (Text Book), Araz courses, Baku, Azerbaijan, 2000

    A university exam preparation text book for high school students in Azerbaijan.

ACKNOWLEDGEMENT

I want very special thanks to my students.

1. Ahmet Akyuz
2. Adrian Demerson
3. Mehmet N Sakarya
4. Nathaniel Taivo
5. Ekrem Ersoy
6. Thy Kalynn Vo
7. Bruce Tannous-Oliveras
8. Kayla Tran
9. Sakura Mikeska.
10. Ahmet Mahir Inalhan
11. Kerem Sarioglu
12. Tiffany Nguyen
13. Parth Shah
14. Reyansh Maheshwari
15. Nozima Agzamova
16. Nathan Nguyen
17. Fadeella Syed
18. Emmanuel Mariga
19. Emily Villeda
20. Angelina le
21. Ihsan Kanli
22. Ronaldo Ivan Martinez
23. Martin Bate
24. Afia Shaik

For your effort and feedback.

Tayyip Oral
555 MATH BOOK SERIES AUTHOR
10/03/2023

www.ingramcontent.com/pod-product-compliance
Lightning Source LLC
Chambersburg PA
CBHW080828220526
45467CB00008B/2235